Optics for Ophthalmologists

Alfredo A. Sadun James D. Brandt

Optics for Ophthalmologists

A Board-Review Manual

With 56 Illustrations

Springer-Verlag
New York Berlin Heidelberg
London Paris Tokyo

Alfredo A. Sadun
Departments of Ophthalmology and Neurological Surgery
University of Southern California
Estelle Doheny Eye Institute
Los Angeles, California 90033, USA

James D. Brandt
Department of Ophthalmology
Los Angeles County/USC Medical Center
University of Southern California
Estelle Doheny Eye Institute
Los Angeles, California 90033, USA

Library of Congress Cataloging-in-Publication Data
Sadun, Alfredo A.
Optics for ophthalmologists.
1. Eye—Accomodation and refraction—Outlines, syllabi, etc. 2. Optics, Physiological—Outlines, syllabi, etc. I. Brandt, James D. II. Title. [DNLM: 1. Optics—examination questions. 2. Reflection, Ocular—examination questions. WW 18 S1260]
RE925.S23 1987 612'.84 87-23357

Reprint of the original edition 1987

Text prepared by the editors using a McIntosh Plus computer system with an Apple Laserwriter printer and a Microsoft Word 3.0 word processor.

9 8 7 6 5 4 3 2 1

ISBN-13: 978-0-387-96623-6 e-ISBN-13: 978-1-4612-4810-1
DOI: 10.1007/978-1-4612-4810-1

for Debra and Karen

Preface

This manual is written to assist the student of optics or the ophthalmology resident in reviewing optics. It is particularly suitable as an optics Board review. However, it was not intended to accomplish the far greater task of encompassing the entire field of geometric optics or even of teaching all that the ophthalmologist needs to know of visual optics or refraction. This manual represents the distillation of lecture notes for an optics Board-review course given at the Massachusetts Eye and Ear Infirmary, Harvard Medical School, for the last seven years. This optics lecture series, which has also been given at Boston University and at the University of Southern California, has traditionally taken place over a week, with three two-hour sessions. Obviously, this six-hour lecture series could not possibly cover the entire field of optics for clinicians. This optics Board-review lecture series has customarily involved a few problems at the end of each lecture which then provided a basis for a review that was given at the beginning of the next lecture. These problems have always proven a useful means of self-testing as well as practice for becoming more comfortable and facile with the concepts involved. The problems have not been taken from the American Board of Ophthalmology test materials, either written or oral. However, they incorporate many of the concepts and computational manipulations which are frequently tested on the ophthalmology boards. At the end of this book, we have provided 24 multiple choice questions of the type that one might expect to encounter on the written American Board of Ophthalmology test. We recommend that the reader hold off on tackling the multiple choice questions, and instead work through the book from beginning to end, solving the problems at the end of each chapter. We believe that this can be accomplished in three or four evenings, and the reader will be rewarded with a more comfortable understanding of optics.

Finally, we must give credit to the pioneers whose teaching approach in optics has provided the basis for the optics Board-review lecture series and this manual. We are particularly indebted to Paul Boeder, who has taught visual optics to several generations of ophthalmologists through the Harvard Optics Lecture Series, given both in Boston and as part of the Lancaster Course in Maine. We would also like to extend our thanks to Melvin Rubin, whose textbook Optics for Clinicians (Triad Publishing, 1974) introduced several teaching approaches that allow the novice to approach the field of geometric optics in a far more intuitive manner. For those who wish to read a book on optics, we strongly suggest Dr. Rubin's book Optics for Clinicians, as well as the companion piece, The Fine Art of Prescribing Glasses (Without Making a Spectacle of Yourself) by Benjamin Milder and Melvin Rubin (Triad Publishing, 1978). A more comprehensive and weighty text on optics and vision is Visual Optics and Refraction — A Clinical Approach, by David D. Michaels (C.V. Mosby, 1980). We hope that this manual, though limited in scope, will help the ophthalmology resident or other student of optics feel more comfortable in dealing with many issues of geometric optics and refraction.

Alfredo A. Sadun, M.D., Ph.D.
James D. Brandt, M.D.

Table of Contents

Chapter I — Principles

Chapter II — Refraction

Chapter III — Applications

Chapter I — Principles

The science of optics can be divided into physical and geometric optics. Physical optics is of greater interest to physicists; geometric optics is far more important for those interested in refraction and the instruments of refraction. Insofar as the eye is an instrument of refraction, ophthalmologists use principles of geometric optics in understanding the interactions of light with the optical system of the eye, prior to the transduction of light in the retina.

Physical Optics

Light as a particle. Many phenomena of the eye can be most easily understood by regarding each photon of light as a particle. Historically, this was a popular perspective of light, especially after it was demonstrated that light could travel through a vacuum and it was shown that the amount of energy transmitted by a photon was quantal. Each particular photon carries with it a certain amount of energy related to its wavelength.

Light as waves. A second point of view is that light exists in the form of a wave; visible light represents a certain range along the electromagnetic spectrum. Indeed, light possesses many properties of waves. Its wavelength and frequency are related to the speed of light and inversely related to each other ($c = \lambda \cdot \nu$). The speed of light varies with the medium which it travels through. However, since the energy is related to frequency (ν) by Planck's constant ($E = h \cdot \nu$), then it follows that at slower speeds (as light travels through different media), it is λ (the wavelength) that changes, and not the frequency.

Huygen's Wavelets. One excellent demonstration of the wave-like properties of light was made by Huygen. Huygen's wavelets also demonstrate the properties of diffraction and interference. As noted in the figure (1) below, light, like any other waveform, treats an aperture through which it must travel as a new point source from which rays emanate in all directions.

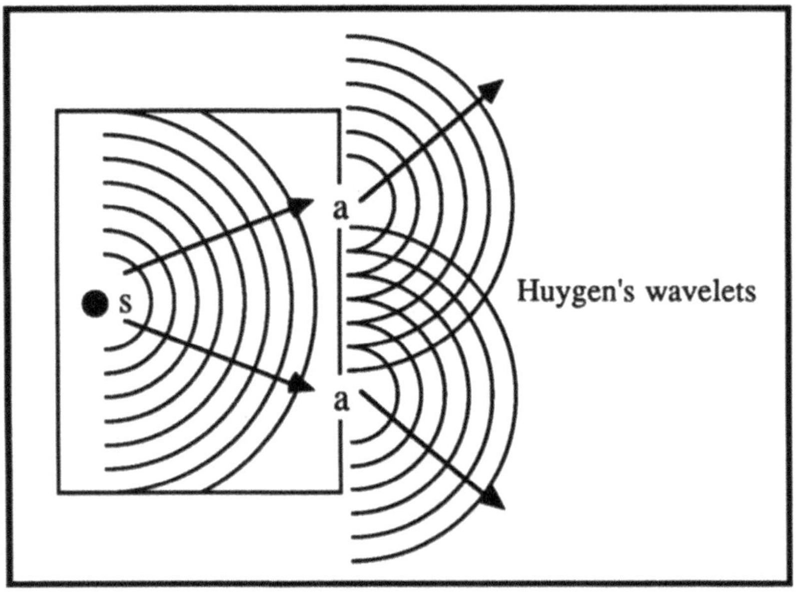

Figure 1 Waves emanate from a source, (**s**). However, they don't travel "through" an aperture (**a**) but, instead, the apertures act as new point sources.

Thus, apertures or even straight edges (Fig. 2) produce sinusoidal variances in the light transmitted through or by them.

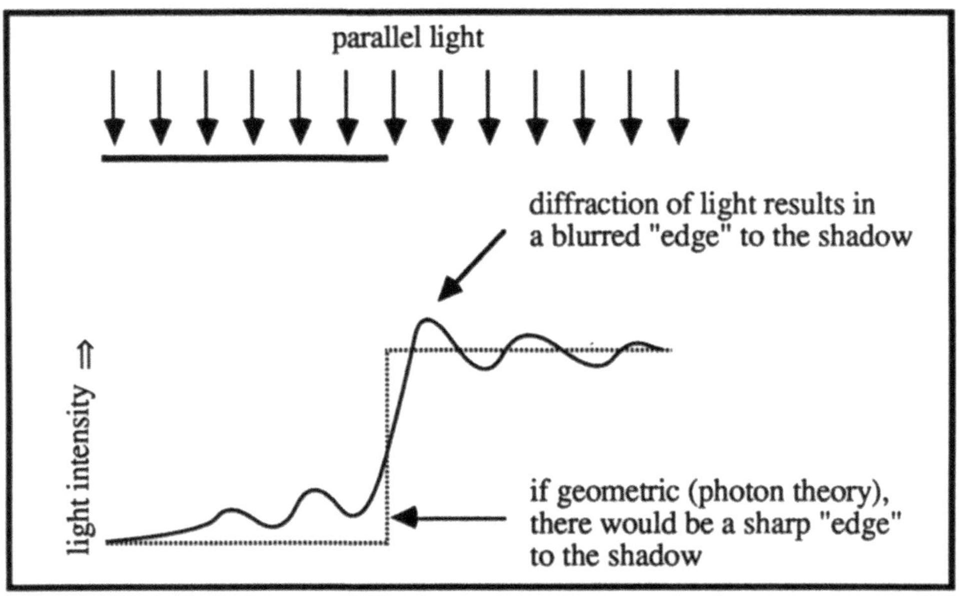

Figure 2 A "knife edge" blocking light does not produce a sharp shadow edge. Diffraction occurs at the knife edge as if it were partly acting as an aperture.

Diffraction is described by the relationship S (in radians) = 1.22 λ/a. As seen in the next figure, light travelling through an aperture (a) produces peaks and valleys of intensity related to the size of the aperture. If the aperture were infinitely small, then light would appear to be emanating from it as a point source.

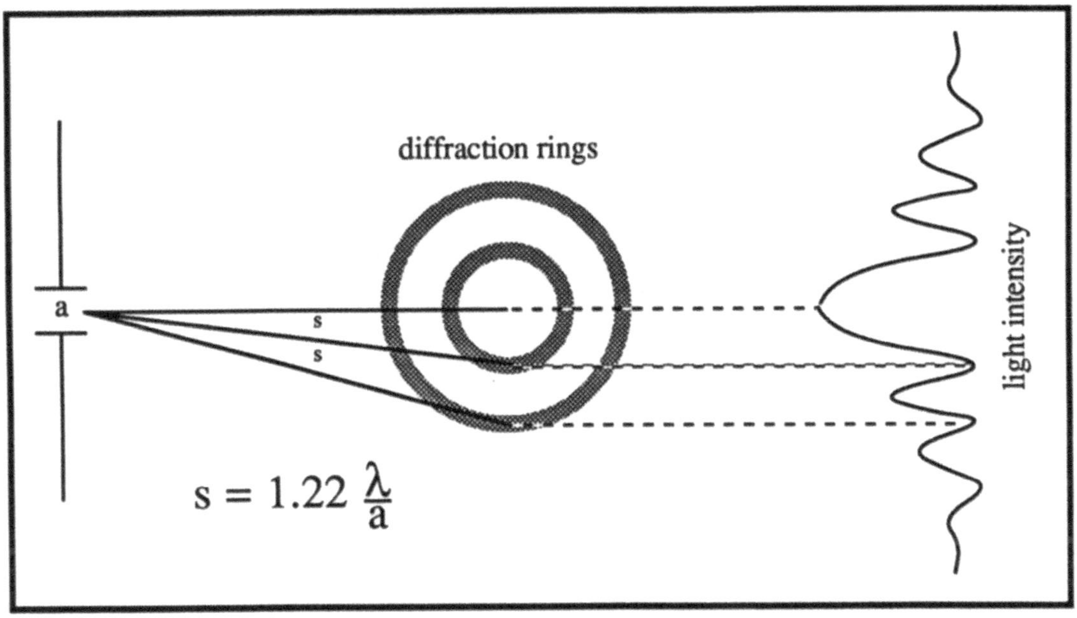

Figure 3 Diffraction rings of light and dark are separated by angle **s**, which is directly related to wavelength (λ) and inversely related to aperture size (**a**).

Interference describes the relationship of two sets of waves, each with its own peaks and troughs. The waves may add their peaks or there may be cancellation as one trough overlays the other wave. One practical application of this principle is in the optical coatings of lenses (Fig. 4).

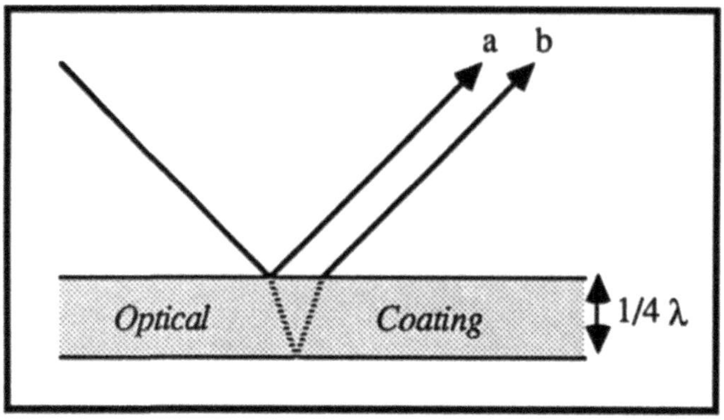

Figure 4 Ray "b" travels 2 • 1/4 λ, or 1/2 λ, and thus is out of phase with ray "a". Optical coatings create two reflected rays that are one-half wavelength out of phase, thus causing mutual cancellation of the (undesired) light with wavelength λ.

Scattering is a third phenomenon of light that is significant to the student of optics. Scattering occurs when light interacts with the elements of the media through which it passes; scattering is inversely related to the wavelength of light, *i.e.*, the shorter the wavelength, the more scattering of light that occurs. Thus the "Tindall blue effect" is the explanation for the blue sky and the yellow appearance of the sun. The blue component of sunlight is scattered the most, fills the sky (which would otherwise be black) and leaves the sun as more yellow (white minus blue). The more atmosphere the light must traverse, the greater the effect. Thus the explanation for the red sunset – at dawn or dusk, light from the sun must traverse more atmosphere. Blue light, having the shortest wavelength, undergoes the greatest amount of scattering, and thus is subtracted from our direct view of the sun and instead fills the atmosphere with indirect light. The least scattered (and refracted) red rays of the sun remain to be seen in orange or red sunsets.

Geometric Optics

By assigning a single line, perpendicular to the wavefront of light, one can simplify the interactions of light with refracting or reflecting surfaces. This is regarded as the study of *geometric optics* and provides a valuable approach to the study of the human eye. Henceforth, since the ophthalmologist generally works in dimensions much greater than the wavelength of visible light, we will depict light as a single ray, ignoring both its particle and wave-like characteristics. However, we will need to explain certain definitions and conventions used in studying light rays.

Definitions

Vergence

Vergence represents the direction of the ray of light perpendicular to a wavefront. This can be represented in a drawing by a single straight line. *Divergence* is mathematically represented by minus numbers; in nature, light is always diverging. *Convergence* is represented by plus numbers. Man-made objects can create convergence. Convergence of light to a single point creates an *image point*.

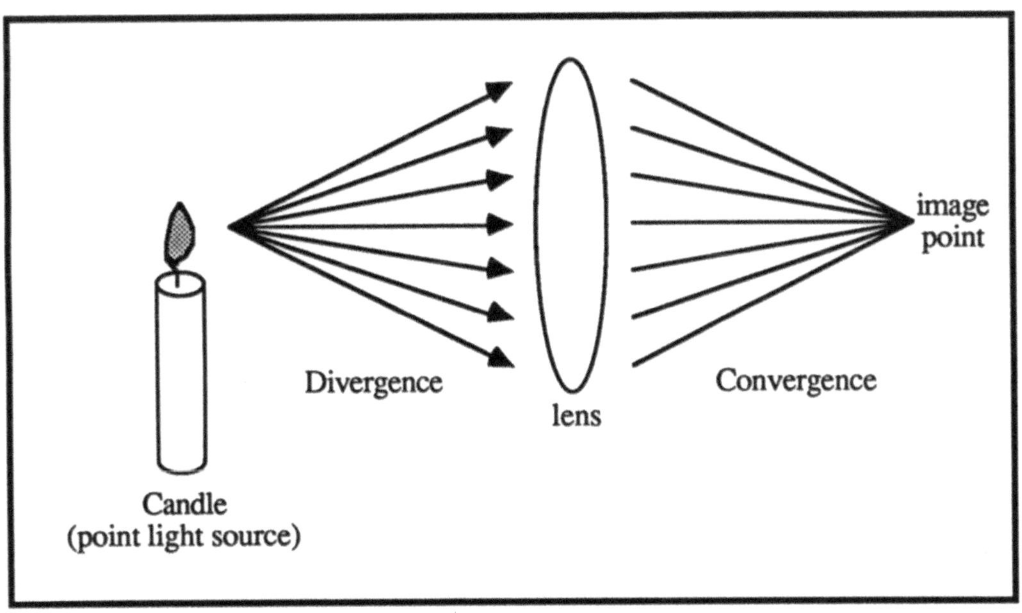

Figure 5 Light in nature diverges. Only certain lenses and mirrors can produce convergence.

Diopter

The units of divergence or convergence are *diopters*. One diopter is defined as 1/distance of the rays of light from an object in meters (D=1/n). Therefore, the greater the distance from the point source, the smaller the divergence value. Similarly, the greater the distance from the point to convergence, the smaller the convergence value. The sun represents an object point which is so far away that it can be regarded as infinite in distance. It therefore has a divergence value of 1/infinity, which equals 0. Zero divergence (or zero convergence) describes parallel lines. We regard sunlight as parallel rays of light.

Images

Images may be *real* or *virtual*. Real images represent the actual points of convergence of light. For example, a projector forms a real image on a screen. Virtual images are images which that appear to be in a certain place through refraction or reflection of light, yet the light does not physically emanate from these image points. An example of a real image is the retinal image seen in the hand lens of an indirect ophthalmoscope; an example of a virtual image is the view seen through a direct ophthalmoscope.

Object Space vs. Image Space

Rays of light leave an object point in object space. After transversing through a lens and having their vergence changed, they enter image space and may come to converge to an image point (Fig. 6). If they do not converge to a real image point, they may diverge, creating the impression of a virtual image from which they would appear to have emanated – such a virtual image would be in front (to the left) of the lens.

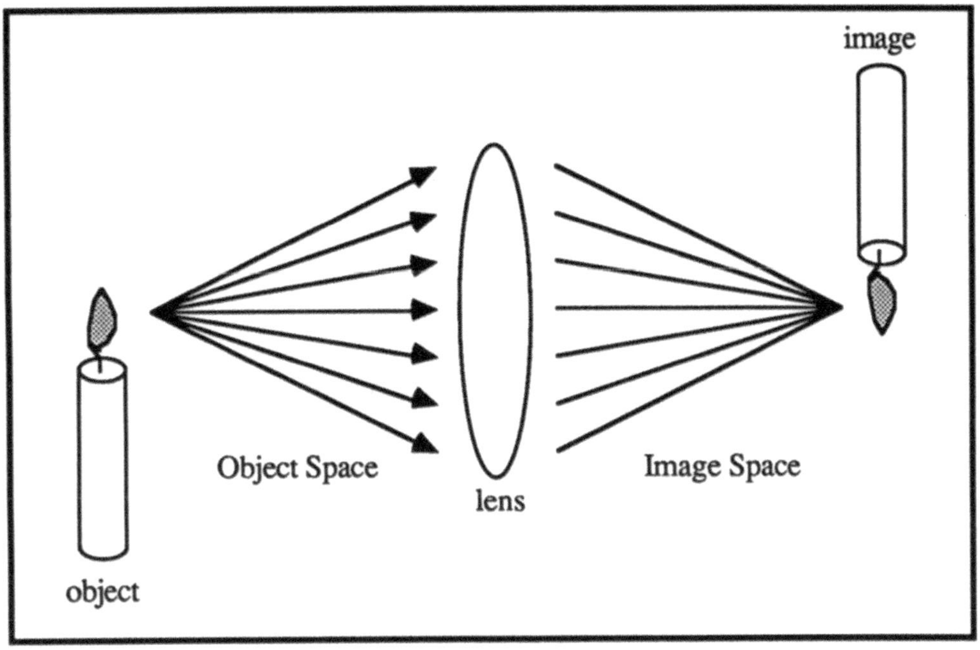

Figure 6 Light leaves an object in object space, acquires a new vergence by passing through a lens and enters image space. It may converge to a "real image" or diverge from an apparent or "virtual image".

This brings us to certain conventions used in describing the above concepts.

Conventions

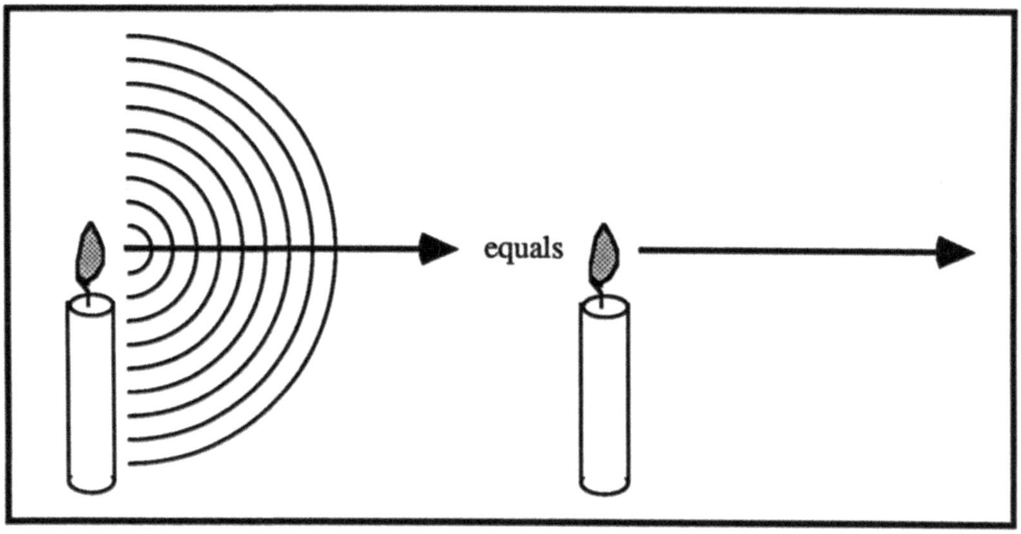

Figure 7 A straight line depicts the direction of light.

• Light is always described as beginning on the left and proceeding to the right. A straight line is used to represent a ray that is perpendicular to the wavefront of light.

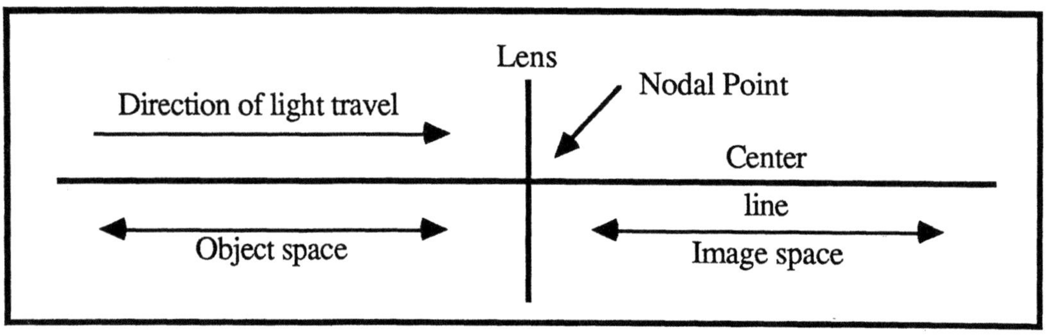

Figure 8 Light passes through a lens at any level. Only if it passes through the nodal point does it continue in a straight line, unaffected.

• Light impinges on a lens from left to right. The lens is regarded to be infinitely thin and infinitely long. The object space is therefore to the left of the lens and the image space to the right of the lens. A second lens, however, requires a redefinition of object and image space.

• A center line exists that goes through the nodal point of all lens systems. The *nodal point* of a lens is the only portion of a lens through which a ray of light will travel and not be refracted.

• The ophthalmologist regards any distance greater than 6 meters (about 20 feet) as being so far away as to be considered infinite. Similarly, light impinging on a lens from a point source greater than 6 meters away is considered to be parallel.

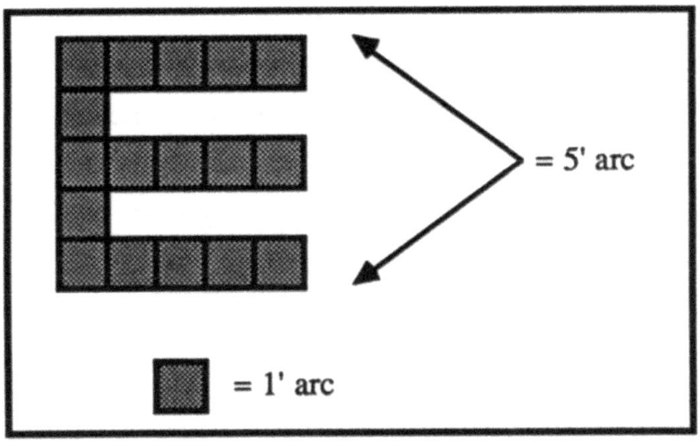

Figure 9 Each component of a 20/20 letter subtends one minute of arc; therefore the entire letter stands five minutes high. At twenty feet, this is about 0.9 cm.

• When describing visual acuity, we use *Snellen letters*. A 20/20 letter is that which is usually seen at a distance of 20 feet, but not necessarily. It is defined as being composed of lines one minute of arc in size. On the average, each letter requires 5 parts and is therefore 5 minutes of arc in height.

The Relationship of Objects, Lenses and Vergences

$$U + P = V$$

In this fundamental equation of geometric optics, **U** represents the vergence of the object, **P**, the power of the lens and **V**, the vergence of the image. All units are in diopters.

In the simplified state of light travelling through air, the index of refraction is approximately 1, and this simple equation stands by itself:

$$U = \frac{1}{u} \text{ (u is given in meters)}$$

$$V = \frac{1}{v} \text{ (v is given in meters)}$$

For example, as seen in figure 10, U is an object 50 centimeters to the left of a lens of a power +6. Small u has a value of .50; Large U = -2 (-, since light diverges from a point). The lens has a power of + 6 (being +, this is a converging lens).

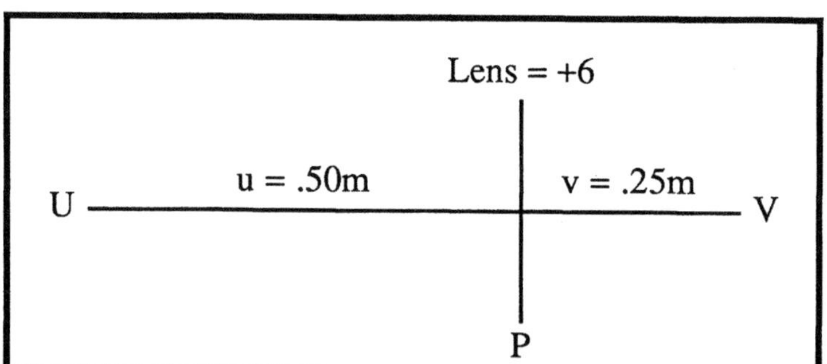

Figure 10 Light leaves the object with a divergence of -2, gains +6 convergence at the lens and finally, with a convergence of +4, forms an image 0.25 meters to the right of the lens.

$$U + P = V$$
$$-2 + 6 = +4$$

Thus, V has a value of +4 ; 1/v = 0.25 (in meters). The image point must be 25 centimeters to the right of the lens.

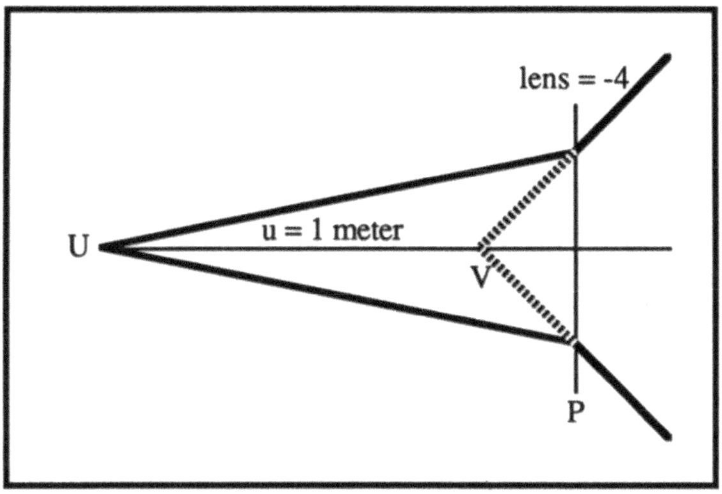

Figure 11 Light diverging from the object point gets an extra amount of divergence from the lens. The "virtual image" is to the left of the lens.

Figure 11 shows another example using a minus lens. Here, u is 1 meter. Thus, the value of U must be -1 (diverging, remember). The power of the lens is -4. U + P = V ⇒ -1 + (-4) = -5. If V is -5, then v = 1/-5, or -0.20, which is **minus** 20 centimeters.

On which side of the lens should it go? Being of minus value, this time V is to the *left* of the lens system. However, as you can see from the drawing, light does not converge to V to form a real image as it did in figure 10. Rather, light diverged from U and the lens gave it a further "kick" of divergence so that now it diverges even more. This produces a *virtual* image, V, and not a real image, V. To an onlooker on the right who is unaware of the lens system, the impression would be that light is simply coming from point V located 20 centimeters behind the lens.

Remember the equation, U + P = V; it is the fundamental equation of geometric optics.

The Nature of Lenses

We have described P in the U + P = V equation as giving the power of a lens in diopters. What is a lens? A lens is a transparent medium that refracts light. Most simply, one can imagine two types of lenses, double convex or double concave. The former cause the convergence of light (and thus have a + value for our equation); the latter produce a divergence of light (and thus have a - value for P).

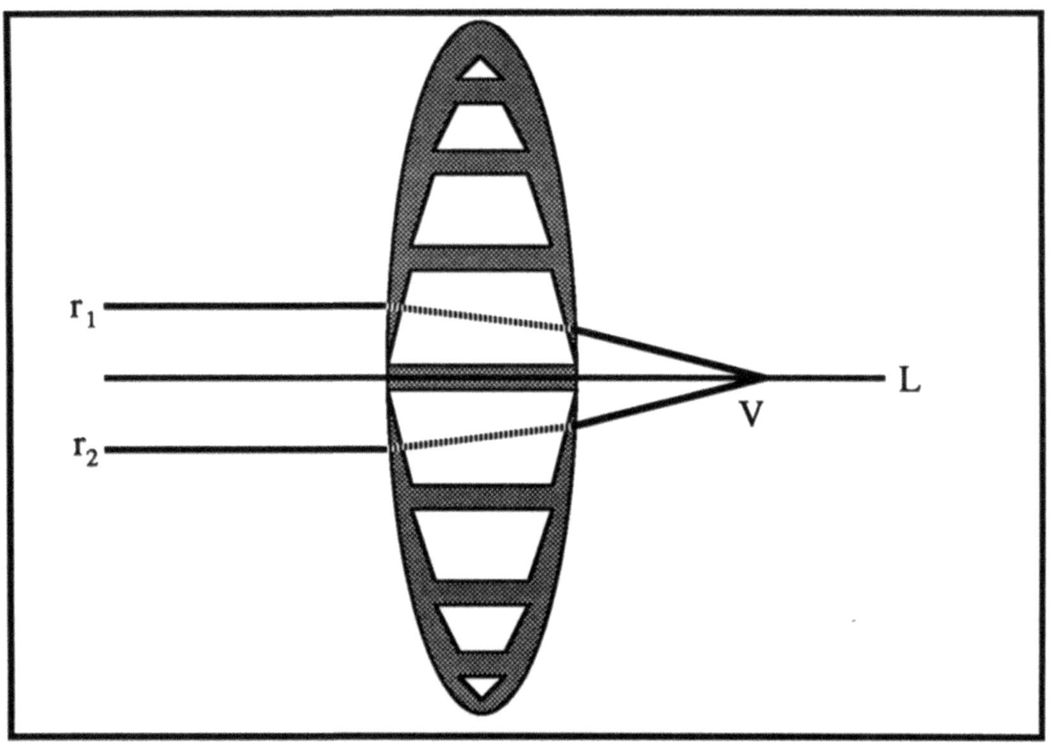

Figure 12 A "plus" lens visualized as stacks of prisms with their bases facing the middle of the lens.

As can be seen in figure 12, a double convergence lens can be visualized as simply a multitude of stacked prisms. All of the prisms above line L have their broader base down and their apex up. Each of the prisms is slightly more steeply sided as you move away from the line through the middle of the lens. Each of the prisms below the line has its broader base up and the apex down.

As you know, light striking the side of a prism is bent towards the base of the prism. Therefore, light striking the prisms above the line is bent down; light striking the prisms below the line is bent up. If all the prisms were of equal power, then light above the line would cross and pass below the line while light below the line would cross and pass above the line, all at different points along the line. However, in a lens, the smooth curvature of the double convex lens represents slight changes in the power of an infinite number of prisms so that all of the light converges to a single point somewhere along line L. Thus, the two rays of light, R_1 and R_2, converge at point V along line L due to the plus power of this lens.

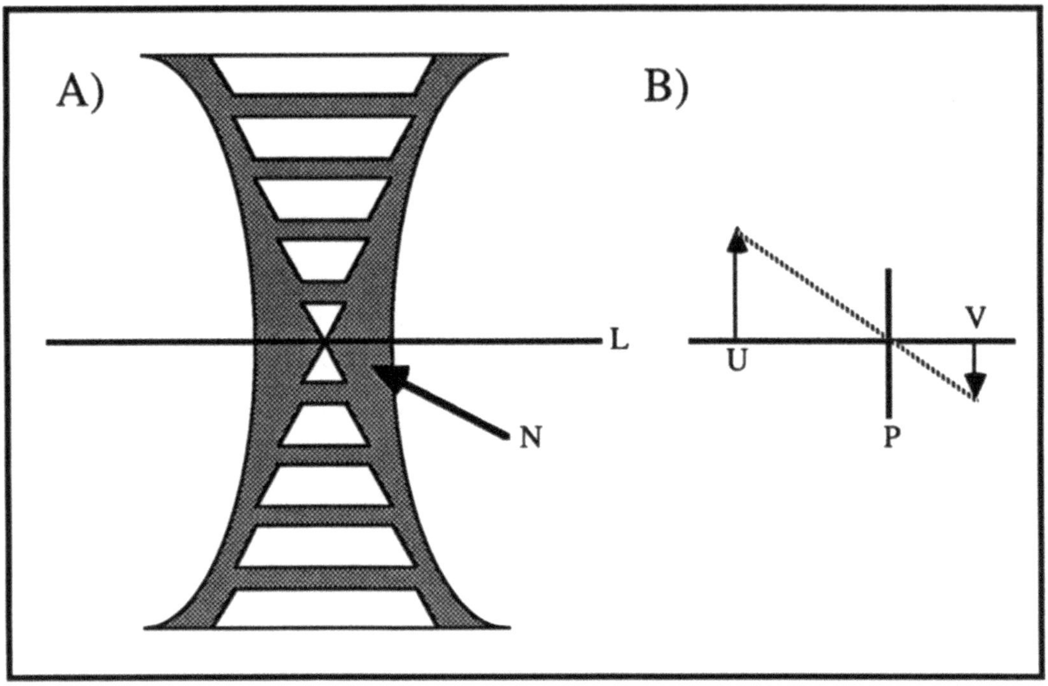

Figure 13 A "minus" lens can be visualized as a stack of prisms with bases facing out (A). Only light rays passing through the nodal point (N) are not refracted. This is shown schematically in (B).

Similarly, a double concave lens can be visualized as stacks of prisms with their bases away from the middle. Again, since light is bent towards the base of a prism, light striking above the midline L will be refracted even higher, whereas light striking below this line will be refracted lower. This minus (double concave) prism is therefore going to produce *divergence.* Note, however, that light travelling right through the middle along line L will not be refracted upward or downward. Indeed, there exists a point right through the middle of the lens called the nodal point, through which a ray of light can pass in a straight line. Light that passes through the nodal point remains undeviated.

In the ideal thin lens system of geometric optics, the nodal point is said to be where the lens and the center line cross. So as you see in figure 13B, regardless of the power of the lens P, the object U can have rays drawn through the center of P (nodal point) that go undeviated onto the image of V.

When light strikes a lens anywhere other than the nodal point, it becomes refracted. The physical principles of refraction will be dealt with shortly. However, at this point it will be useful to get an intuitive grasp of the concept. When light comes to an interface that separates two media (each with a different index of refraction), it becomes deviated. This is a function of the change in its velocity between each medium.

Several analogies can be made to physical objects which, when they change speed, deviate their line of course. One common such occurrence is when a four-wheeled object such as a toy wagon rolls down a smoothly paved sidewalk. Its course may be relatively straight. However, when one of the wheels catches the grass alongside the sidewalk, that wheel slows down. This change in speed causes the wagon to turn *towards* the grass alongside the sidewalk. In other words, a moving object will turn *into* the medium that slows its velocity.

A slightly more accurate analogy can be envisioned as follows: Imagine, as in the figure below, several lines of soldiers marching along a well paved, straight road. As the road turns, the soldiers on the rightmost ranks step out into the grass. This slows down their pace, causing them to drop back relative to the rest of the file. Given the constraints of straight files, the ranks of soldiers will be forced to march onto the grassy field at a different angle from that which they maintained while on the paved road (which permitted faster marching). As the figure shows, the deviation of the line of march is again *into* the direction of the medium that reduces the speed of travel.

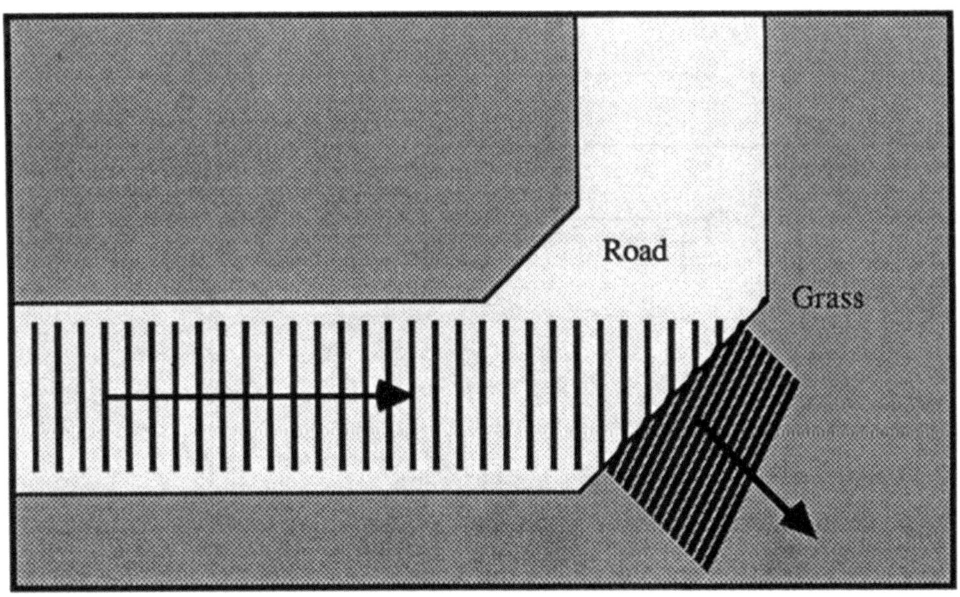

Figure 14 Soldiers in marching files change the direction of their march when they go from marching on a high speed road to the slower, off-road grass, yet their files remain straight.

Lenses and Their Focal Points

As discussed above, a "plus" (+) lens can be considered as a series of prisms with the base in the middle. Such a lens adds convergence to light striking it. If the final convergence is greater than 0, then a plus lens will bring light to focus on a real image.

A "minus" (-) lens can be considered, as above, to be a stack of prisms with their apices oriented towards the center. It adds divergence to light striking it. Unless light comes to a minus lens already converging more than the power of the lens, then light passing through will not come to focus at a real image point.

The focal point of a lens can be considered to be a measure of the lens power. F_1 is that point from which light, if it were to diverge and strike the lens, would then emanate as parallel light (Fig. 15A). In other words, F_1 represents U in the equation $U + P = 0$ (remember, a vergence of 0 results in parallel light and produces no real or virtual image). F_2 is the reverse. Imagine parallel light striking the lens; F_2 represents the distance to the image point created by the lens in its action on parallel light (Fig. 15B).

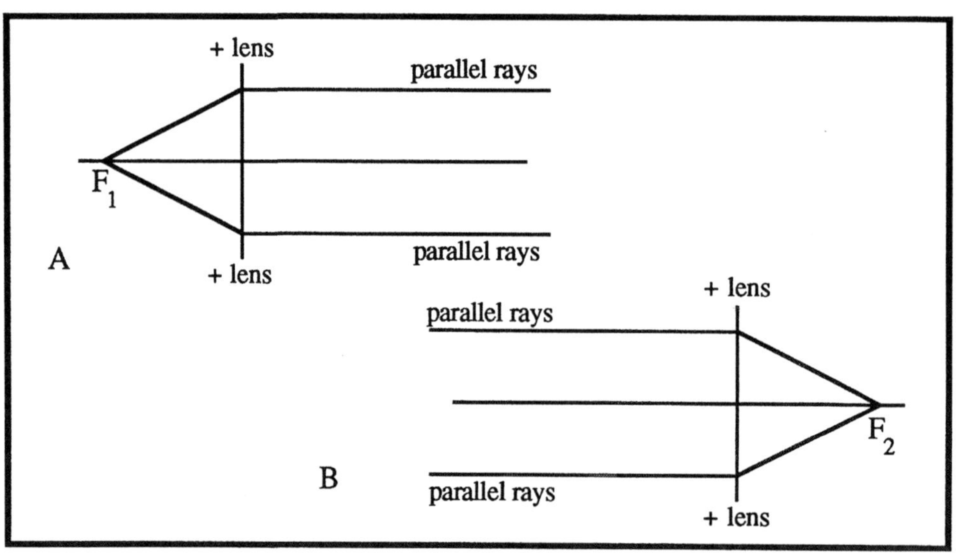

Figure 15 F_1 is that point which produces divergence which, in combination with the lens is zero (parallel light). F_2 is the image point produced when parallel light strikes the lens.

Generally, when one describes the focal length of a lens, one is describing the distance between the lens and F_1.

Pushing Points

Rubin has popularized the exercise of pushing points to familiarize the student of optics with the effect of the equation U + P = V in several settings. This exercise is particularly useful because it gives an intuitive grasp of the effects of lenses on light from objects with various vergences. It also helps in appreciating the principle that parallel light can be conceived as converging at an infinite distance in the image space or diverging from an infinite distance in the object space.

Consider Figure 16 below. The object O_1 is located 1 meter to the left of the +4 lens. It therefore has a divergence of -1 ($U + P = V \Rightarrow -1 + 4 = +3$) and light converges 33 centimeters to the right of the lens at I_1. If we push this point closer to the lens, as in O_2, we're now only 50 centimeters in front of the lens; the divergence is -2 ($U + P = V \Rightarrow -2 + 4 = +2$) and the light converges at I_2, 50 centimeters to the right of the lens. Pushing the object point further to the right, to position O_3, brings it 25 centimeters before the lens. The divergence value is now -4 ($U + P = V \Rightarrow -4 + 4 = 0$); there is no image point because the light leaves the lens as parallel rays. Finally, pushing the object point further to the right, to position O_4, brings it 20 centimeters from the lens. It now has a divergence value of -5 ($U + P = V \Rightarrow -5 + 4 = -1$); image I_4 is a *virtual* image 1 meter to the *left* of the lens, existing at the same point as O_1. In other words, as we pushed the object point from left to right, the image point also moved from left to right until it went off the end at infinity (parallel light) and then returned at the other end, on the left again. Further pushing of the object point to the right would result in virtual image points, also moving to the right, but always to the left of the object point.

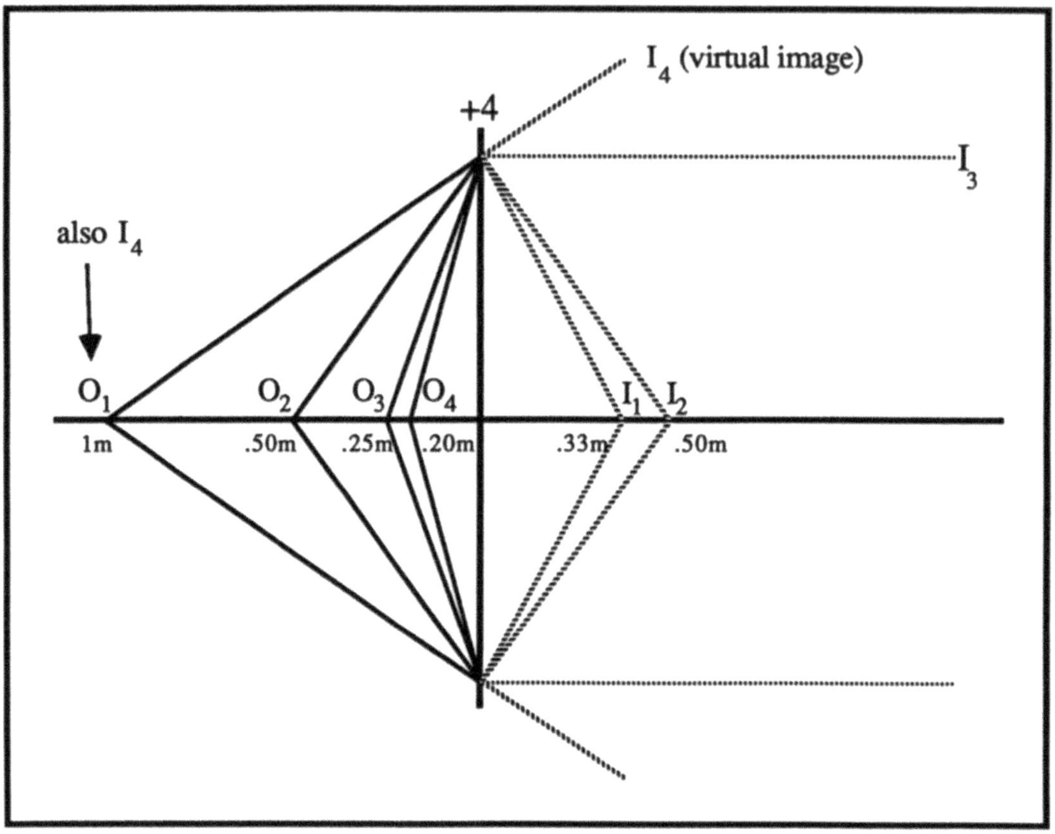

Figure 16 "Plus" lens: Pushing the object points from O_1 to O_2 to O_3 to O_4 causes the points to first move to the right (I_1 to I_2), then to infinity right or left (I_3), then again to the right (I_4), but always behind (to the left of) the object point (O_4).

A similar exercise in pushing points can be made with the minus lens. In Figure 17 below, object 1 (O_1) is located 1 meter to the left of the lens; hence it is diverging -1 ($U + P = V \Rightarrow -1 - 4 = -5$). Divergence is therefore increased to -5 and the image point is a virtual image 20 centimeters to the left of the lens. O_2 is located 50 centimeters to the left of the lens, and hence has a divergence of -2 ($U + P = V \Rightarrow -2 - 4 = -6$); the virtual image is therefore one sixth of a meter, or approximately 16.7 centimeters to the left of the lens. It is clear that the further one moves the object point to the right the further the virtual image point also moves to the right, always staying to the right of the object point but to the left of the lens.

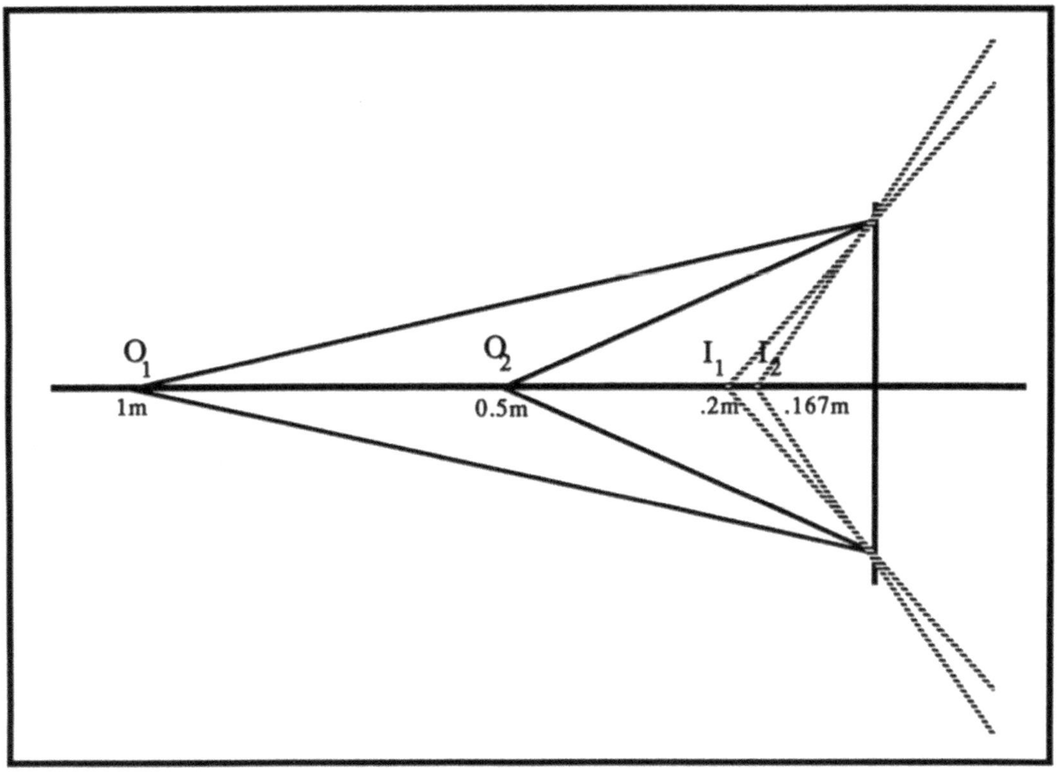

Figure 17 "Minus" lens: Pushing the object point to the right always pushes the image point further to the right.

Multiple Lens Systems

When dealing with multiple lenses, the same conventions and definitions as described above apply. U + P = V is still the crucial equation. When dealing with the multiple lens system, one simply applies the equation sequentially. The object point of the second lens is simply the image point of the first.

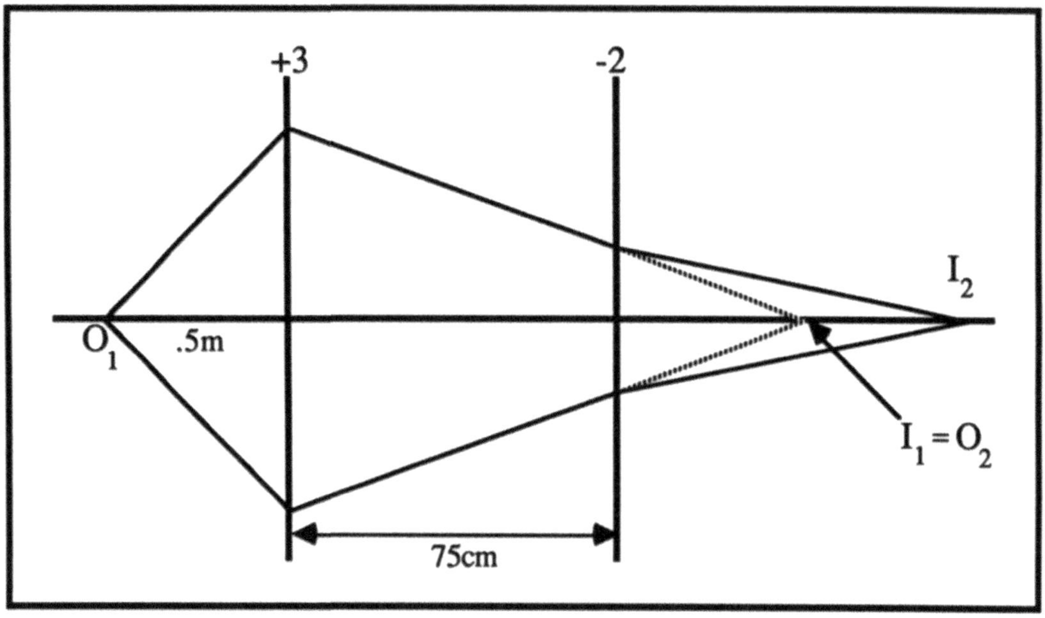

Figure 18 An object point (O_1) is located 0.5 m to the left of the first of two lenses separated by 0.75 m. The image point of the first lens becomes the object point of the second lens ($I_1 = O_2$), and the calculation is re-performed.

For example, see Figure 18 above. An object point (O_1) exists 50 centimeters to the left of the first lens, which is a +3 diopter power lens. 75 centimeters to the right of the +3 lens is a -2 lens. In solving for the final image point, one simply applies the equation $U + P = V$ to the first lens, and then uses that image point as the object point of the second lens. Beginning with the first object point (O_1), which is .5 meters to the left of the first lens and hence has a divergence of -2 ($U + P = V \Rightarrow -2 + 3 = +1$), the resultant image is a real image point (I_1) located 1 meter to the right of the +3 lens. Having solved for I_1, one now calls that object point O_2. O_2 is converging to a point 1 meter to the right of the first lens, and hence 25 centimeters to the right of the second lens (1 m - .75 m = .25 m). Its convergence in relationship to the second lens is therefore 1/0.25 = +4. The new object point vergence (U) equals +4. We now apply the equation $U + P = V$ again and get +4 - 2 = +2. The second image point (I_2) is therefore a real image 50 centimeters to the right of the second lens. Clearly, this process could be reapplied for as many lenses as one wished in a multiple lens system.

Magnification

$$M = \frac{U}{V}$$

Magnification is best described by the relationship M = U/V. This is to say that magnification is directly proportional to the vergence power of the object, and inversely proportional to the vergence power of the image for any lens system. When the value of M is positive, it means that the image remains upright. When the value of M is negative, it means the image is reversed. When the value of M is greater than 1, M describes the proportion of magnification. For example, if M is 4, there would be a magnification of 4 times the original size. When the value of M is less than 1, then M describes the proportion of minification.

For example, see Figure 19. An object point 50 centimeters to the left of a +6 lens produces a divergence of -2 ($U + P = V \Rightarrow -2 + 6 = +4$). A real image is formed 25 cm to the right of the lens. The magnification equals U/V or -2/+4, or -0.5. The image formed by such a lens would therefore be upside down and half the size of the original image.

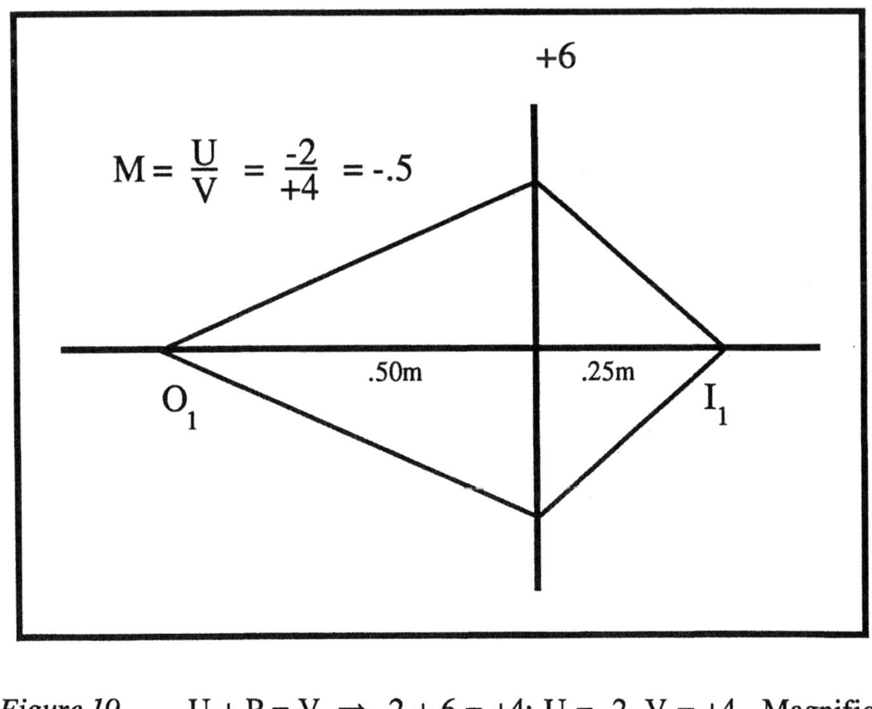

Figure 19 $U + P = V \Rightarrow -2 + 6 = +4$; U = -2, V = +4. Magnification (M) equals -0.5, and so the image is half size and inverted.

Gullstrand's Eye

Professor Allvar Gullstrand (1862-1930) was an ophthalmologist who, through a study of postmortem dissections as well as clinical evaluations of living human eyes, came up with a schematic model of the human eye. This schematic eye established lengths and optical properties that produced refractive properties very similar to those of the living human eye. Using his model, he was able to locate optical cardinal points, which permitted very useful geometric manipulations. Gullstrand is the only ophthalmologist ever to receive the Nobel prize. It was given for "work on the diffraction of light by lenses applied to the eye"; his most noteworthy contribution to clinical ophthalmology was the development of the modern slit lamp.

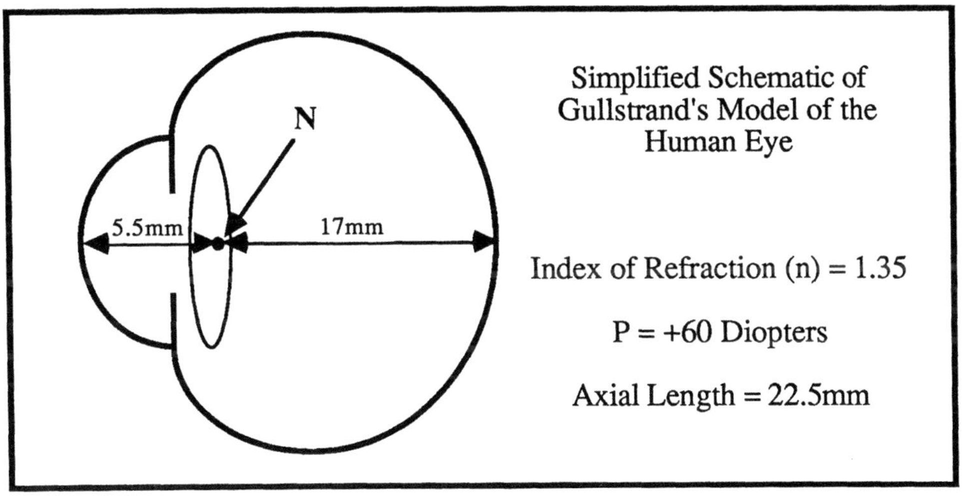

Figure 20 Note that the nodal point (**N**) is 5.5 millimeters from the cornea and 17 mm from the retina.

As seen in Figure 20 above, a simplified form of Gullstrand's schematic eye is based on a combined lens power (which in the human eye would be composed mostly of cornea and lens) of +60, which has a nodal point 5.5 mm behind the front of the eye. It is then 17 mm further back to the retina. The media of the eye are all averaged out to have an overall index of refraction (n) of 1.35. This schematic eye has an axial length of 22.5 mm.

Some of the geometric manipulations that can be performed with Gullstrand's eye will be addressed later in this manual. At this point though it is worthwhile bringing up the eye for several intuitive definitions of clinical syndromes:

Emmetropia • In the emmetropic eye, parallel light (travelling from a distance) strikes the front of the eye and comes to focus 22.5 mm back on the retina. When light emanates from an object close to the eye (*e.g.*, 1 meter away), it has a divergent power (*e.g.*, -1), which requires the eye to change its lens power in a corresponding amount. In such an example, the lens would change its shape to increase the total power of the eye from +60 to +61. As a result, the light rays would still come to focus on the retina.

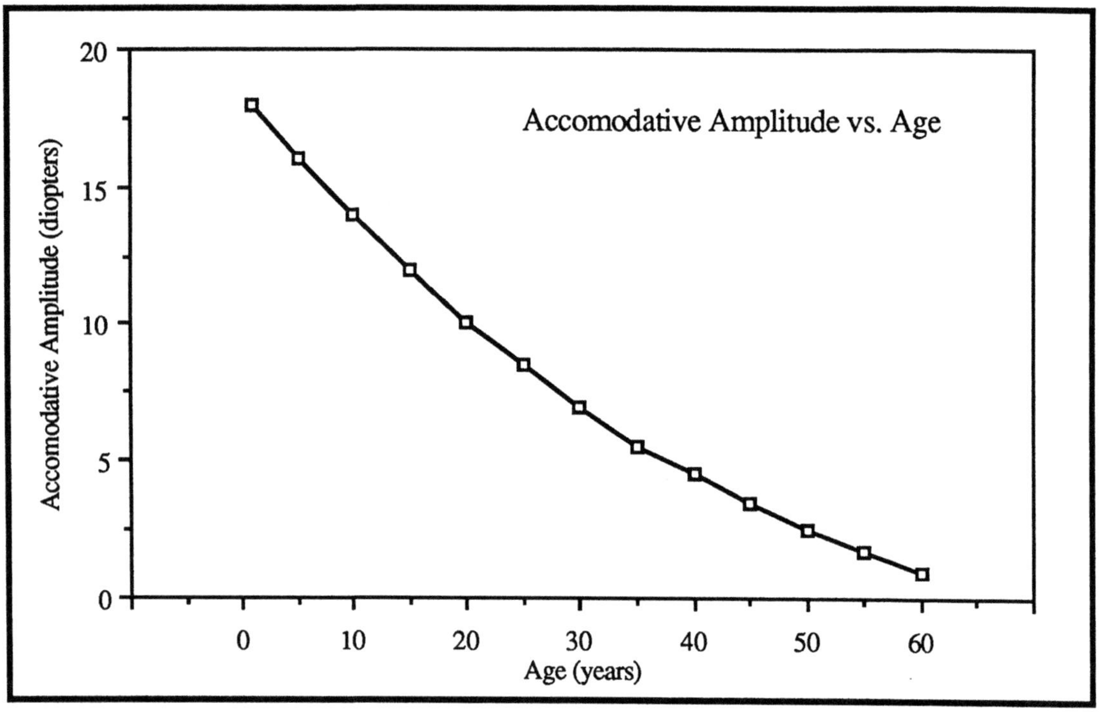

Figure 21 With age, accommodation decreases. The emmetropic eye requires assistance ("plus" lenses) to see at near (14 to 18 inches) past the age of about 45 to 50 (**Presbyopia**).

Accommodation/Presbyopia • The process by which the lens can change its shape to increase its power in order to overcome the divergence of objects at near is called accommodation. The very elastic eye of children is capable of a great deal of accommodation, but with age this capacity diminishes. The graph above shows the decreasing capacity for accommodation with age. Ultimately, there are insufficient reserves of accommodation available to neutralize the divergence of light coming from objects kept at the usual distance for reading or manipulation (usually about 40 centimeters), which would require about 2.5 diopters of accommodation. At this point, usually at age 45 or 50, the individual will need reading glasses in order to see clearly at near. Note that the 45 year old patient still has 3.5 diopters of accommodation available to his lens; however, all this accommodation cannot be used for protracted periods of time. Therefore, a person of this age finds it far more comfortable to use only about half of his accommodative capacity (approximately 1.75 diopters) and to make up the remainder with reading glasses. It should be obvious that once a patient has reached the age of presbyopia and uses such reading glasses, that they will not see clearly at distance with the use of these glasses since additional convergence is now being applied to light arriving from a distance (and is thus coming in roughly parallel).

Myopia • Later on a more useful definition and description of myopia will be given. However, for the moment, simply consider the Gullstrand eye with either P (the total lens power) being too great (>60) or the axial length being too long (>22.5 mm). In either situation, light striking the refractive components of the eye will come to focus in front of the retina instead of on the retina. Such light will then produce a blur-circle instead of a precise point on the retina.

Hyperopia • The hyperopic eye is the opposite of the myopic eye. In the Gullstrand model it can be considered as an eye whose total lens power (P) is less than 60 or whose axial length is shorter than 22.5 mm. In either case, there is not enough convergence power to bring parallel rays at distance to focus on the retina. The light would come to a focal point somewhere behind the retina but, being interrupted at the retina, would produce a blur-circle as well.

Astigmatism • The astigmatic eye is not symmetrical along each axis. Certain axes of the eye are more or less hyperopic or myopic than other axes, producing distortion as well as blurring of the images projected upon the retina. Correction of the astigmatic eye needs to be made with the use of cylindrical lenses; these are lenses whose power is not symmetrical over an entire sphere, but is greater at one axis. More on this later.

Rethinking Myopia and Hyperopia

In the previous section the intuitive definition of myopia was that the eye has either a refractive system that is too strong (resulting in over-convergence of incoming rays to a point in front of the retina) or an axial length that is too long. Though this is intuitively comfortable, there is a better way of defining and describing the myopic eye, and it is this: *The myopic eye has a far point that is close in front of the eye*. Conversely, the hyperopic eye should be thought of not as being too small or having too weak a converging system, but rather as having a far point that is *behind* the eye. What follows are the definitions of far point and near point.

A *far point* is the plane from which light emanates that will produce a sharp image on the retina without there being any accommodation by the eye. Issues of astigmatism are not taken into account.

The *near point* is that plane from which light emanates which would result in an image being focused on the retina with maximum accommodation by the eye; again, astigmatism is not taken into account.

Let us reflect further on the far point. The myopic eye can be considered as a perfectly healthy eye (there is nothing wrong with the strength of its lens or with its axial length) which, however, was designed not to look out at distance without accommodation but, rather, to look at a nearby target. Consider, for example, a -2.5 diopter myopic eye. Light coming from a distance (parallel rays) would focus in front of the retina and therefore produce a blur-circle. However, light emanating from an object 40 centimeters in front of the eye would have a divergence of -2.5 and therefore come to focus exactly on the retina of this myopic eye. Thus, we could define this myopic eye as one that sees perfectly clearly, without accommodation, objects at a distance of 40 centimeters in front of the eye. Indeed, this would be an eye ideally suited for reading since it would never tire from excessive use of accommodation. This is an eye that nature designed for looking at objects 40 centimeters away.

Conversely, a hyperopic eye can be considered to have been designed to put light in focus onto the retina which already comes in with a certain amount of convergence. This means, however, that the far point must be defined as being behind the eye, where the rays of light would come to focus if the eye did not interact with them.

These concepts of far point being in front of the myopic eye and behind the hyperopic eye may seem confusing now, but will be made clear when manipulating vertex distances. These concepts are critically important in being able to comfortably understand the effect of a corrective lens that can be held at various distances in front of the eye to produce different refractive effects. Indeed, the entire concept of a corrective lens can be better appreciated in terms of far point.

Vertex Distances

We start with the understanding that a myopic eye has a far point that is nearer to the eye than infinity. The usual goal in refraction is to optically change this eye to an emmetropic eye. Therefore, the far point needs to be moved optically from its original position to infinity so that the eye can see clearly at distance.

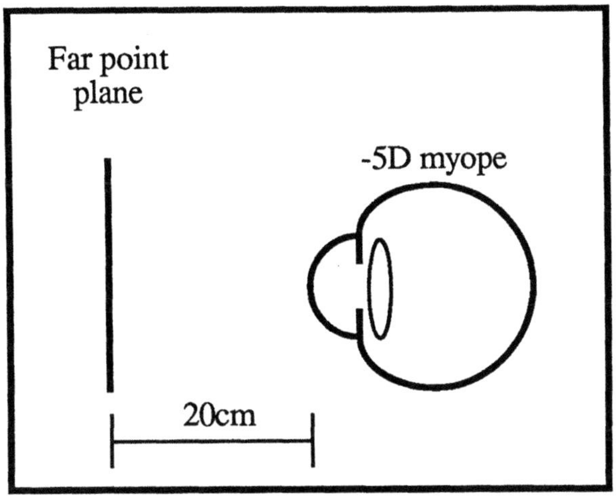

Figure 22 The -5D myope has a far point 20cm in front of the eye. Images in this plane focus on the retina. To see clearly at distance, images must be optically moved to this plane.

For example, consider the 5 diopter myope. He has a far point that is 20 centimeters (1/5 meters) in front of his eye (Fig. 22). He sees everything clearly there. The goal of refraction is to have him see everything clearly at distance (emmetropic). The trick to achieving this, then, is to place a lens of appropriate power and position to change his far point to be that of infinity. Put another way, the corrective lens should optically make infinity appear to the eye as if it were at the plane of the far point. Clearly, a -5 diopter lens at the plane of the eye (*i.e.*, a contact lens) will convert parallel light (coming from a great distance) to light which diverges by a value of -5. Such light will appear to be coming from a virtual image 20 centimeters behind the lens. This places the virtual image at the far point of that myopic eye, thus producing an image that is perfectly in focus on the retina.

What if spectacle correction were used on this same -5 diopter eye? If the spectacles were placed in a plane 2 centimeters in front of the eye, they would now be only

18 centimeters from the far point. If they were to convert parallel light to produce a virtual image at the plane of the far point their power would have to be 1/0.18 or -5.56. You can see that as glasses are placed further away from the eye (*i.e.*, the vertex distance becomes larger), the distance between the glasses and the far point becomes smaller. As this occurs, the glasses need increasing power of divergence (minus diopters) to produce a virtual image at that same far point plane.

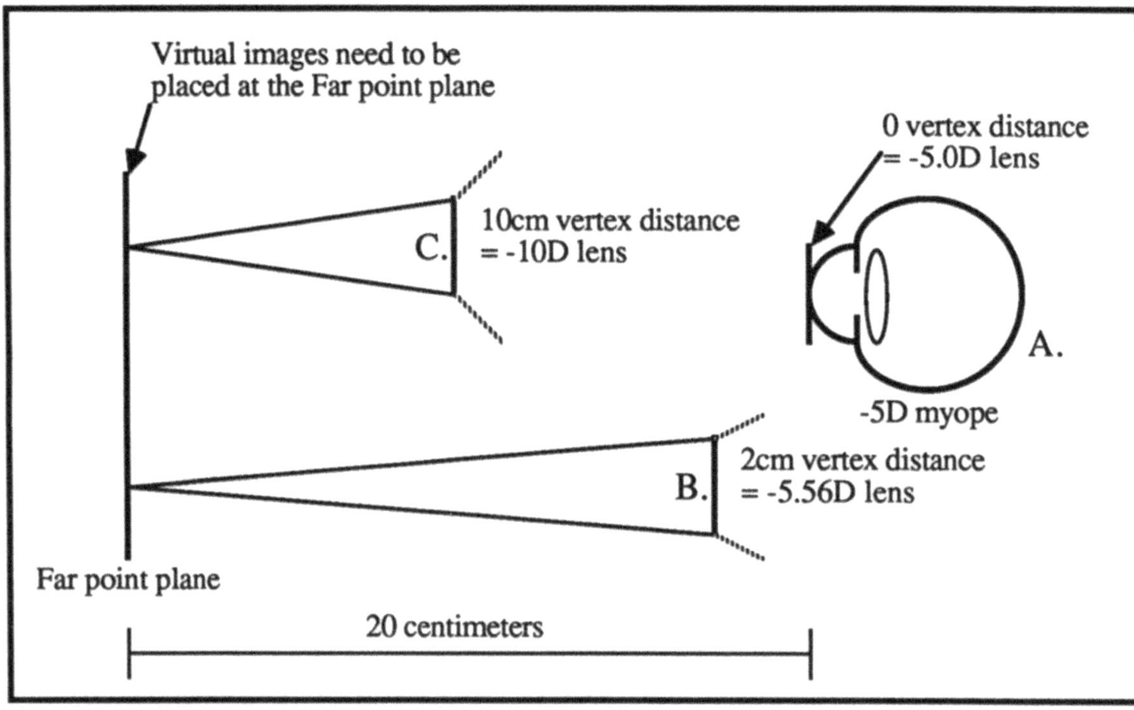

Figure 23 The far point plane is 20 cm in front of the -5D myopic eye (A). Spectacles 2 cm in front of the eye are 18 cm from the plane (-5.56D) (B). Glasses 10 cm in front of the eye (C) are only 10 cm from the plane (-10.0D).

Consider taking this concept to an extreme, as in Figure 23 above. If the glasses were held 10 centimeters in front of the eye, they would then be only 10 centimeters from the far point plane of the -5D myopic eye. The power necessary there to provide a sharp retinal image would have to be the inverse of their distance from the far point. One divided by 0.10 equals 10. Of course, the glasses would have to be divergent in order to convert the parallel light coming from a distance into divergent light from a virtual image at the far point. Thus, the same 5 diopter myope would be adequately corrected by a -5 contact lens, a -5.56 pair of spectacles 2 centimeters in front of his eyes or a -10 diopter lens held 10 centimeters in front of his eyes!

Now it is obvious why the power of a corrective lens needs to be different depending upon its position relative to the eye. When a patient goes from contact lenses to glasses or from spectacle correction at a certain vertex distance to spectacle correction at another vertex distance, then it becomes necessary to apply these principles to recalculate the power of the new set of lenses. In the process of making such a calculation, the *first step is to establish the far point of the eye*. Once the far point of the eye has been identified, then any lens at any distance can be quickly calculated.

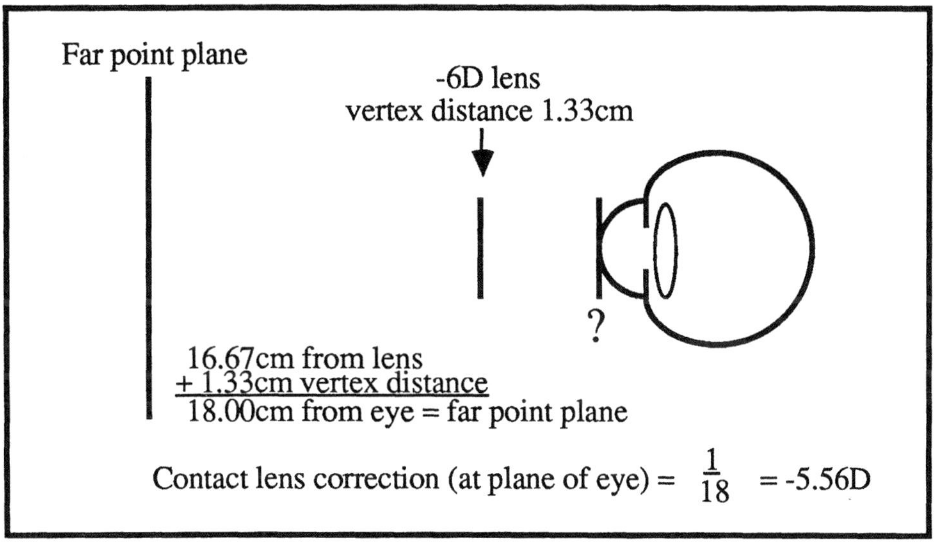

Figure 24 Glasses with a -6D lens, 1.33 cm vertex distance, correct this myope. Therefore the far point is 1/6, or 16.67 cm, to the left of the lens. But this is 18 cm to the left of the eye. Thus the correct contact lens power would be 1/18, or -5.56D.

For example, consider the figure above. In this case a patient wearing -6 diopter glasses wishes to be changed to contact lenses. What power contact lenses should be prescribed? The first step is to establish the far point. In order to do that, we need to know not only that the glasses are -6 diopters, but that the vertex distance (the distance of the glasses to the eye) is 1.33 centimeters. If these glasses at this distance provide a sharp image on the retina, then the far point must be that place where parallel light striking -6 glasses produces a virtual image. 1/6 meters is 16.67 centimeters. Thus, the far point is 16.67 centimeters to the left (in front) of the glasses. However, the far point of the eye is that distance plus the distance between the glasses and the eyes (1.33 centimeters). These two values add up to 18 centimeters. Therefore, this eye is myopic with a far point 18 centimeters in front of it. The rest is simple. By convention, a contact lens is considered to be at the plane of the eye. Such a lens needs to take parallel light and produce a virtual image 18 centimeters in front of the eye. This is simply the reciprocal of 18 and minus (diverging), or -5.56. A -5.56 contact lens will produce the same sharp retinal image when viewing at distance as the original -6 diopter spectacles at a vertex distance of 1.33 centimeters. More will be said later about how to manipulate vertex distance calculations.

The approach to these types of problems becomes less intuitive when dealing with hyperopic eyes. This is because it is difficult to conceive of the far point as being **behind** the eye. Consider the following example: a patient is 4 diopters hyperopic. Intuitively, you might explain this in terms of the patient's lens system being too weak to converge light onto the retina, or the eye being so short that the convergent light doesn't have a chance to come together before it reaches the retina; both situations lead to blurred images on the retina. However, as in the case of the myope, it's more useful to think of the hyperope as having a far point which provides images that come to focus on his retina. Since the hyperope needs light *that is already converging* to strike his eye, such a far point cannot be in front of the eye, for any object there would produce divergent light rays. In fact, the far point of a hyperope is *behind* the eye!

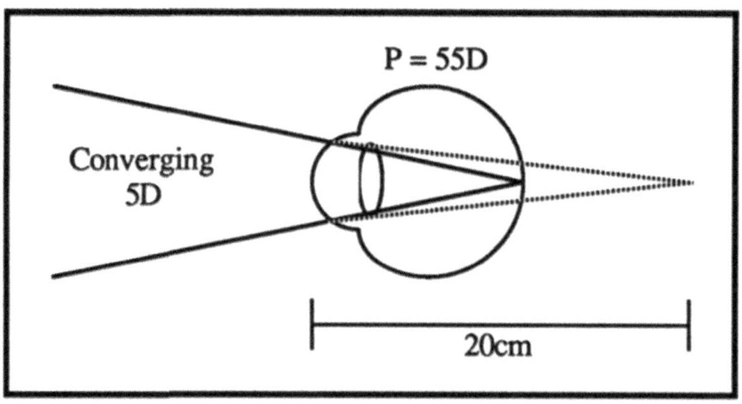

Figure 25 Only light converging to a point 20 cm behind the eye would be focused on the retina of a +5D hyperope.

Consider, as in Figure 25 above, the patient who is 5 diopters hyperopic. Suppose that, compared to Gullstrand's schematic eye, he has the proper axial length but has only +55D as the value of all his convergence lenses. Parallel light therefore would come to focus behind the eye. However, more importantly, if light arrived at the eye with 5 diopters of vergence already, then the 55 diopter power of the lens system would produce 60 diopters of vergence and bring the image to focus on the retina. What would be the object point of such light with that convergence? Light with this convergence would come to focus on a real image 20 centimeters behind the lens system. Although the lens system of the eye has a nodal point 5.5mm from the front of the eye, this is close enough to the front of the eye that, for convenience, we use the front of the eye for our calculations. What lens should then be put on the front of the eye (*i.e.*, a contact lens) to correct this eye's hyperopia? A +5D lens may be put on the cornea:

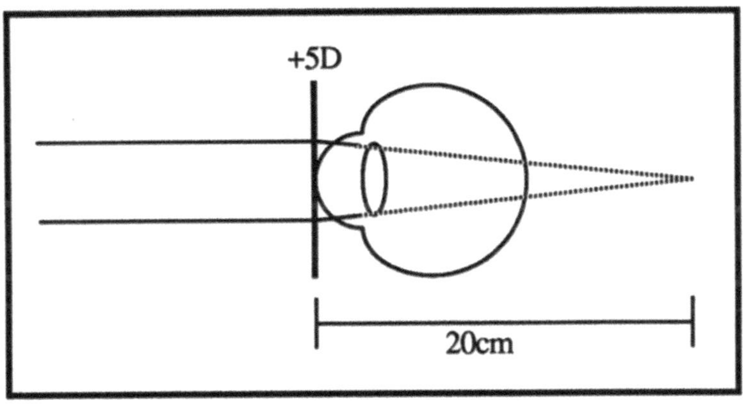

Figure 26 A +5D contact lens makes parallel light converge to the far point of a +5D hyperope, thus allowing images at distance to be in focus.

Such a lens would take parallel light and cause it to converge to a point 20 centimeters to the right. This is at the far point of the hyperopic eye of 5 diopters. Therefore, the hyperopic eye now sees parallel light in focus on the retina.

How would one vary vertex distances with such an eye? For example, supposing this patient develops problems with his contact lens and wishes instead to convert to spectacle correction. What power glasses should he wear if the vertex distance is 1.5 centimeters? This lens, 1.5 centimeters in front of the eye, is therefore 20 + 1.5 cm in front of the far point. Light therefore needs to converge to a point 21.5 centimeters away. The reciprocal of 0.215, which is 4.6, is the value of the correct spectacle correction giving at a vertex distance of 1.5 centimeters, as shown below.

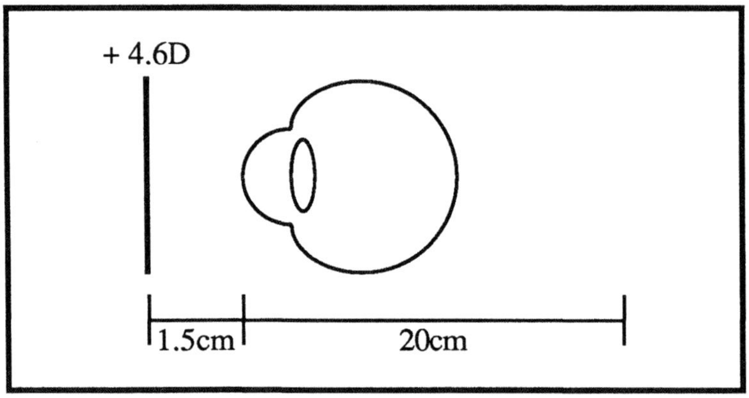

Figure 27 A +5D hyperope can be corrected with a +4.6D lens with a vertex distance of 1.5 cm since the distance from the lens to the far point is 1.5 + 20 = 21.5 cm.

One can see then that as long as the myopic eye is regarded as one that has a far point some place in front of the eye, and the hyperopic eye as one that has a far point behind the eye, different lenses with different powers at different vertex distances can be manipulated. To summarize, the first step is to establish the far point of the eye; the second is to establish, given the vertex distance, the distance between the lens and the far point of the eye. The third step is to take the reciprocal of that distance as the power of the new lens.

Problem Set 1

1. a) Why is the sunset red?

   b) What is chromatic aberration?

2. An object is 25cm before (to the left) of a lens. Describe the resultant image point ("Real" or "Virtual") and its distance from the lens if the lens is:

   a) +1

   b) +3

   c) +4

   d) +5

3. You are looking out a thick (1") window made of plane glass, at an angle of 30°. Would there be a shift in the image if the window were opened? In which direction?

4. Explain

   a) diffraction

   b) interference

5. Explain aperture size and depth of field for either the eye or a camera.

6.

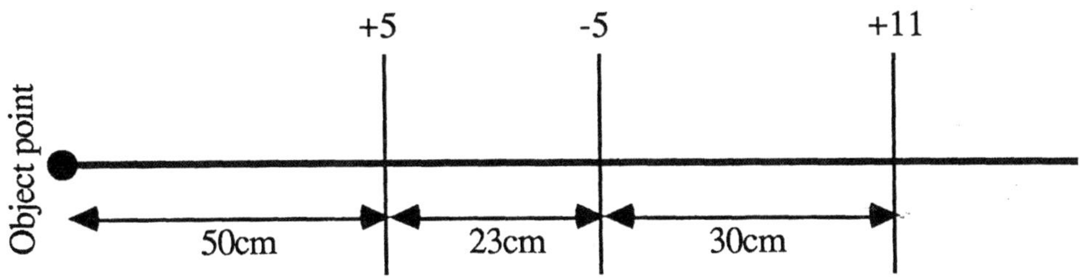

   a) Where are the image points 1, 2 and <u>3</u>?

   b) Is the final image real or virtual?

   c) Is the final image upright or inverted?

7. A light illuminates a Kodachrome slide behind a lens to project an image onto a screen as depicted:

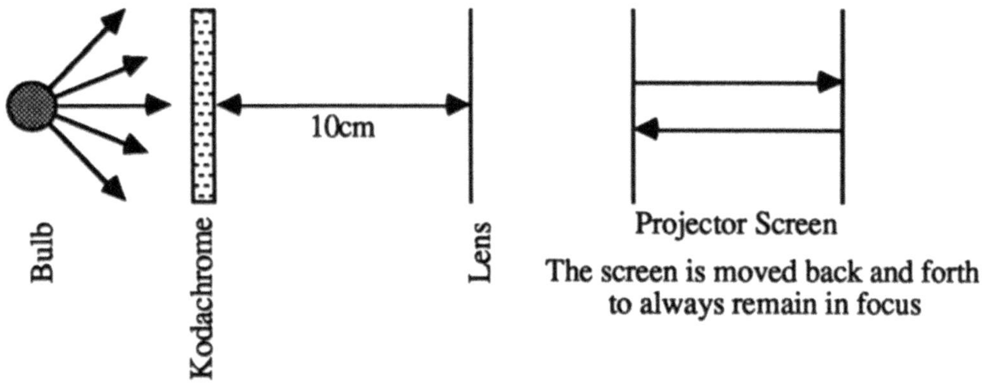

Describe the image, upright, inverted and magnification, if the lens is:

a) +30

b) +2

c) +9

d) -5

Problem Set 1 — Answers

1. a) Shorter wavelengths (*i.e.* blue, violet) are more refracted and more scattered. The remaining light seen is red.

   b) Each wavelength is refracted differently, blue more than red. What is "in focus" for one color will be "out of focus" for any other color. The difference between red and blue refraction is about 1.25D.

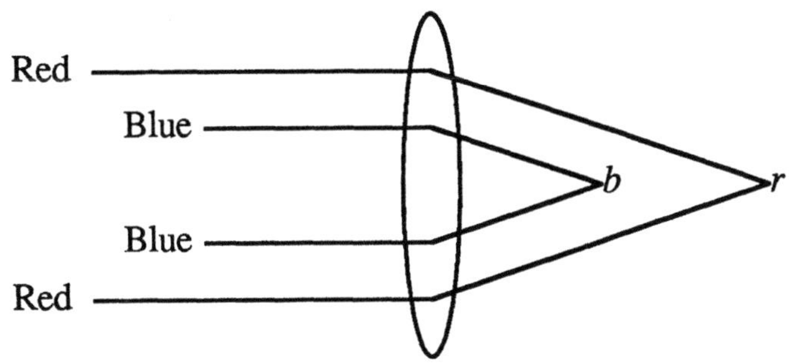

2. **U + P = V**

a)	-4 + 1 = -3	virtual point 33cm	left
b)	-4 + 3 = -1	virtual point 1m	left
c)	-4 + 4 = 0	No image – parallel light	
d)	-4 + 5 = 1	real point 1m	right

3. With the window closed, the image appears as if it is coming from x'. Opening the window permits a direct view of x.

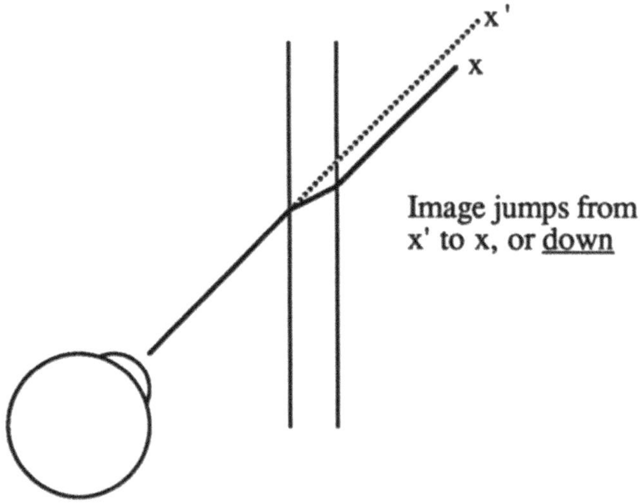

4. a) Huygen's wavelets describe the wave-like character of light. Diffraction occurs as the waves pass by an edge or through an aperture (see Figures 1 and 2).
   b) 1/4 λ film "cancels out" reflections by putting about one half of the reflected light out of phase with the other half (see Figure 4).

5. In a myopic eye such as that diagrammed below, the image plane (where the image comes to sharpest focus) is in front of the retina. An image formed by all the light potentially entering the eye (dotted lines) will cast a "blur-circle" on the retina (grey circle). By placing a pinhole in front of the eye, only the central rays are permitted to enter the eye; they cast a smaller (and thus sharper) blur-circle on the retina. This is entirely analogous to the iris diaphragm in a camera lens – the smaller the *f*-stop, the better sharpness and depth of field.

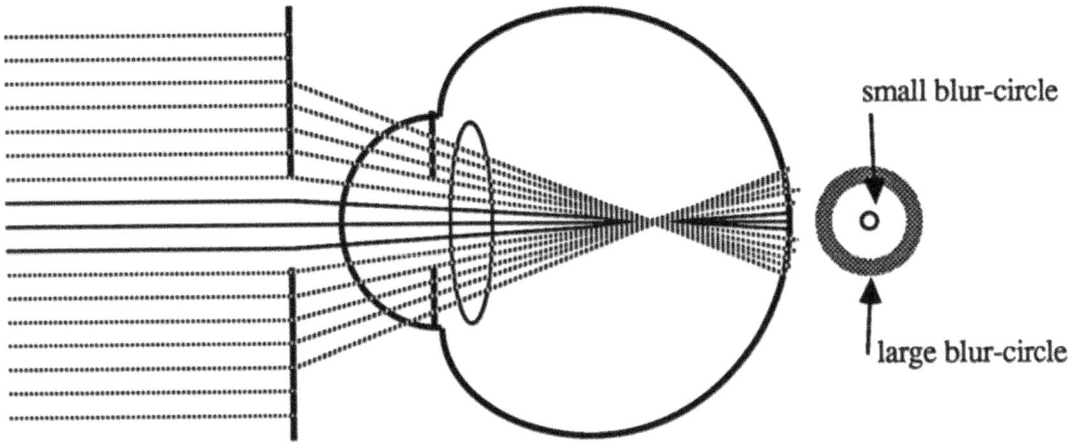

6.

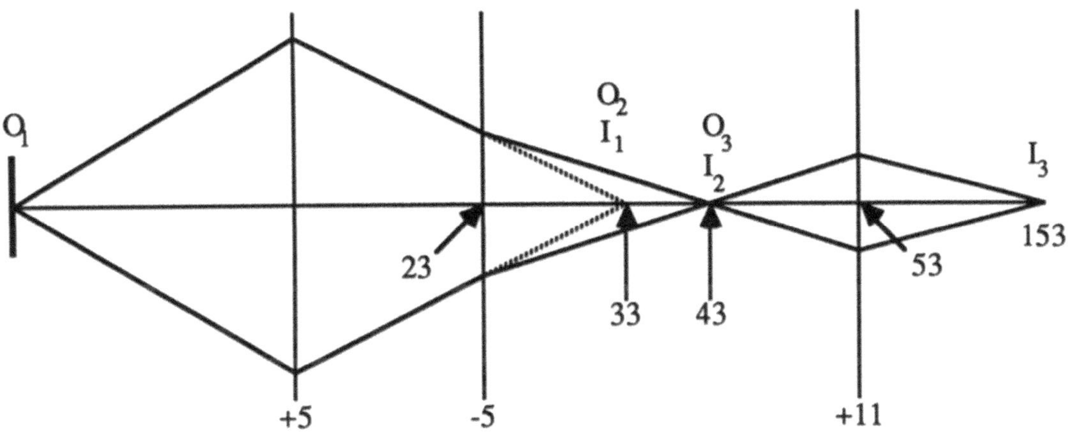

	U + P = V	M = U/V	
O_1:	-2 +5 = +3	- 2/3	a) 33, 43, 153
O_2:	+10 -5 = +5	+ 2	b) real
O_3:	-10 +11 = +1	- 10	c) upright, M = 40/3

7.

	U + P = V	M = U/V		
a)	-10 + 30= +20	-1/2	inverted	minified
b)	-10 + 2 = -8	5/4	upright	magnified
c)	-10 + 9 = 10	10	upright	magnified
d)	-10 - 5 = -15	2/3	upright	minified

Chapter II — Refraction

Points to Planes

Instead of thinking of a single point along the mid axis and describing the vergences in the object and image planes, one can deal with an entire plane. Consider, for example, Figure 28 below. Here we have an object point 1 meter to the left of a +3D lens. However, instead of considering it simply as a point, let us consider the entire plane 1 meter to the left of the lens represented by the arrow. An analysis of U + P = V would suggest that there is a divergence of -1 + 3 = +2, producing an image point 50 centimeters to the right of the lens. Of course, this really represents an image in a plane 50 centimeters to the right. Assessments of this image can be made not only through algebraic calculations, described above, but through a form of graphical analysis which we will now describe.

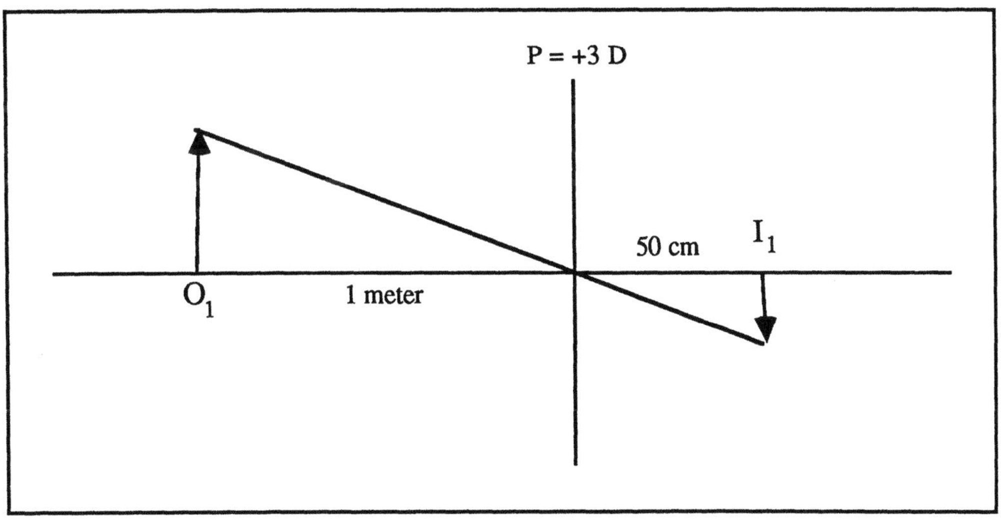

Figure 28 A plane with points at O_1 is refracted by the lens to form a plane at I_1. Note that the ray through the nodal point (the *chief ray*) passes unrefracted.

Graphical Analysis

One can establish a graphical analysis by setting up similar triangles. Towards this end there are 3 rays that can be drawn which will establish the image plane. Actually, one needs only 2 of these 3 rays to find their intersection, but it is useful to be able to manipulate all 3 rays.

The arrow that is shown in Figure 28 extends from the object point up in the same object plane. We needn't worry about the object point since it extends somewhere along the central axis, and the image point will be on the same central axis. Therefore, all 3 rays will emanate and be dealt with in terms of the tip of the arrow, which represents points somewhere else in the object and image planes.

The chief ray begins at the tip of the arrow, but since it passes through the nodal point of the lens, it continues on, unrefracted, in a straight line. The focal point ray is the ray that passes through the F_1 point along the axis for the lens in question. Since any ray from the F_1 will, upon encountering the lens, come out with a vergence of 0 (this is by definition), we now have a second line in the image plane. Finally, the parallel ray begins at the tip of the arrow and proceeds parallel to the center line with a vergence of 0 and strikes the lens. There it is refracted, and by definition, it must pass through the F_2 of the lens and continue in a straight line. The intersection of these 3 rays should form the tip of the arrow in the image plane.

Consider Figure 29. Again, we are dealing with an object that is 1 meter to the left of a +3 lens. For simplicity's sake, the chief ray is not drawn since it is the easiest to visualize. It goes from the tip of one arrow through the nodal point and ends up undeviated at the tip of the second arrow. The second ray is the focal point ray. It starts at the tip of the object arrow and passes through a point along the axis 33 centimeters to the left of the lens. This is the F_1 of the lens and therefore once it reaches the lens it then, with vergence of 0 (parallel), proceeds further to the right. The third ray is a parallel ray. It proceeds with the vergence of 0 from the tip of the object arrow until it encounters the lens. It must then pass through the F_2 of the lens, which is 33 centimeters to the right of the lens. Here it intersects with the focal point ray. This must represent the tip of the image arrow with the base of the image arrow being the image point 50 centimeters to the right of the lens (calculated algebraically).

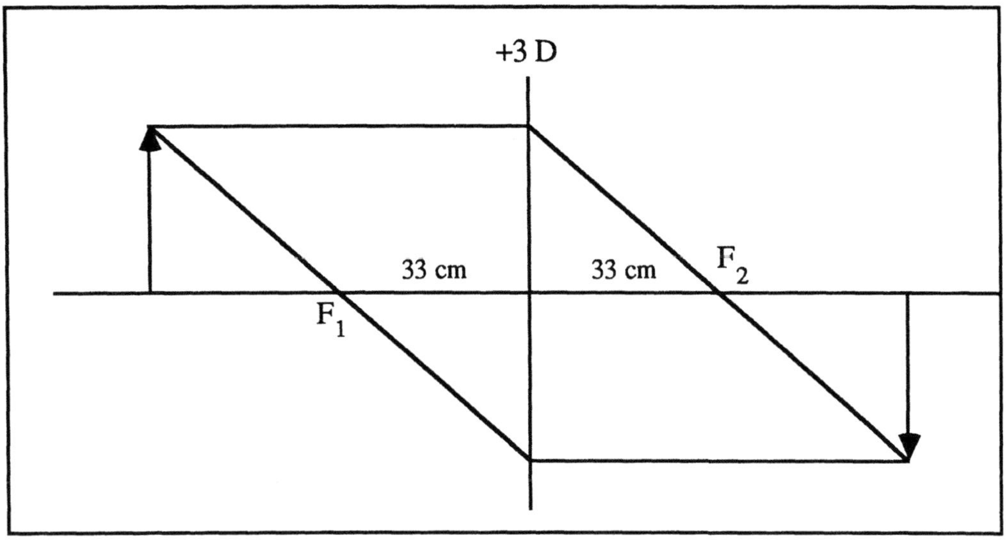

Figure 29 The focal point ray, passing through F_1, leaves the lens with a vergence of 0. The parallel ray is refracted by the lens to pass through F_2. These two rays (and the chief ray passing through the nodal point) meet at the tip of the arrow in the image plane.

Graphical analysis can be extremely useful and will be alluded to later in taking advantage of the known dimensions of Gullstrand's schematic eye. In figure 30, we have only drawn in the chief ray (the one which passes undeviated through the nodal point) having established the object plane and image plane one also has created two similar triangles. The angle θ is the same as the angle θ'. In these two similar triangles, the relative legs a, b and c correspond to a', b' and c'. Naturally, the ratio of a' to a (in the previous figure 0.5:1.0) is the same as the ratio of b' to b, or one-half. Classical trigonometric manipulations can be made from these similar triangles.

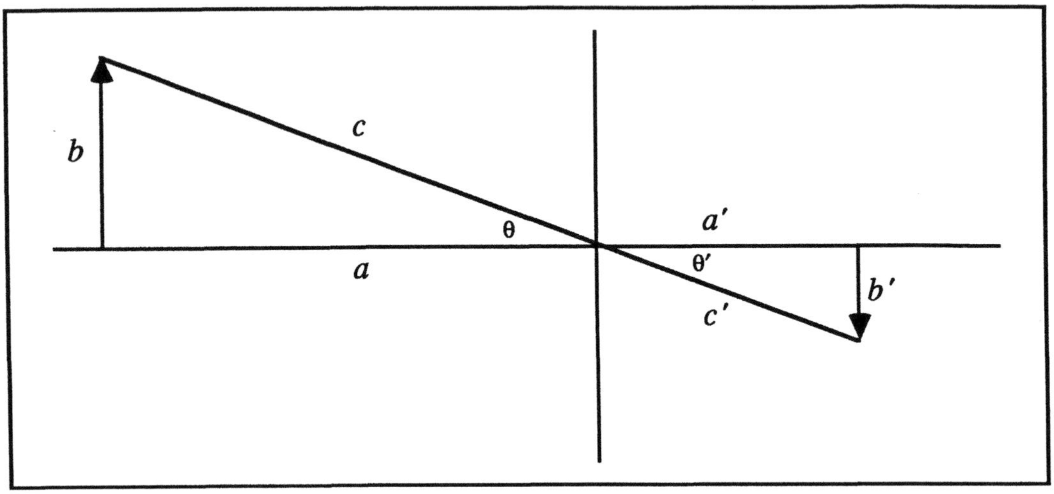

Figure 30 Similar triangles are formed by the chief ray.

Prisms

As noted in the first chapter, lenses can be considered as the accumulation of many prisms of different powers. Let us now discuss a prism in itself. A prism is a 3-sided object whose index of refraction is greater than that of the surrounding medium. Classically, prisms are made of glass or plastic and are used in air. They are held in such a position that light enters parallel to the base of the prism. As it strikes the left hand side of the prism (see Figure 31), it is refracted into the orthogonal direction of the prism, which has a higher index of refraction than does the surrounding air. As light leaves the back side of the prism it is again refracted away from the orthogonal as it goes from a higher to a lower index of refraction. You will notice that both of these deviations of the light are towards the base of the prism. Therefore, each half of the two sizes of prism create deviation of light in the direction of the base. The sum of these two is considered to be the power of the prism. This is defined in diopters and in terms of how much light deviation is created. If a prism deviates light 10 centimeters at a distance of 1 meter, it is said to have a power of 10 prism diopters, written as 10^{Δ}D. It is the ratio of deviation per distance that matters. Therefore, a 10 centimeter deviation at a distance of 2 meters is only about a 5^{Δ}D prism. Similarly, a 20^{Δ}D prism will produce a deviation of 40 centimeters at 2 meters.

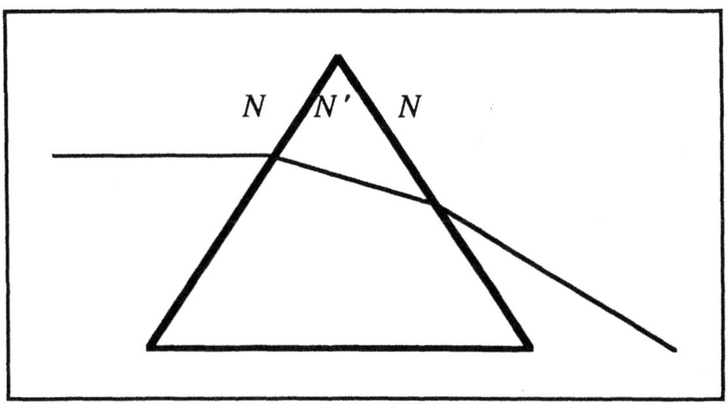

Figure 31 Light is deviated towards the base of the prism as it enters and exits the surfaces of the prism.

Note that although the light is deviated towards the base of the prism, this produces the impression of an image to be deviated towards the apex of the prism. To see this clearly, look at Figure 32 below. The eye sees object x after it has been deviated towards the base of the prism. Of course, the eye believes light to travel in a straight line, and therefore the impression is that x is in position x'. Thus, although light is deviated towards the base of a prism, a prism always creates the appearance of an image deviated in the direction of its apex.

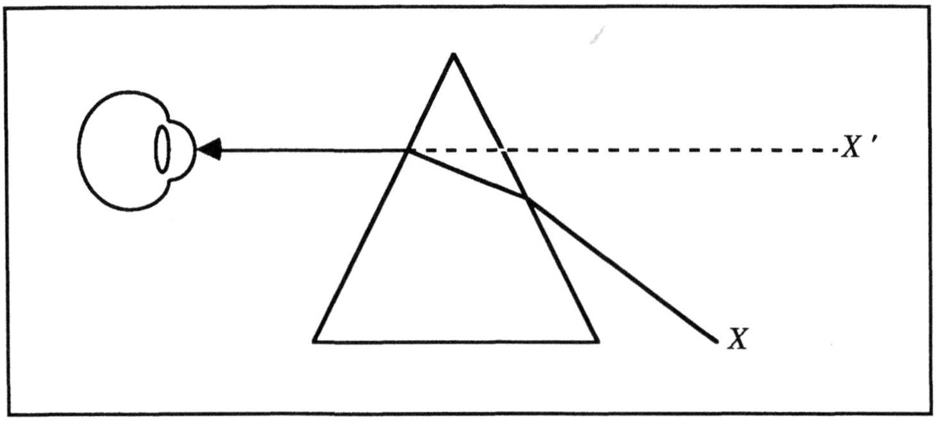

Figure 32 Light is deviated towards the base of the prism. This makes the image appear to be displaced towards the apex.

The relationship of prism diopters (Δ) to angles measured in degrees is that of the tangent of θ. Obviously, this is not a linear relationship and is complicated to calculate each time. For 1°, the tangent of θ is .017, meaning that there is a 1.7^{Δ} change. For 10°, the tangent of θ is 0.176, equivalent to 17.6 as the number of prism diopters. However, for 44 degrees, the tangent of θ is 0.966 which would represent 96.6^{Δ}. So it is apparent that for small angles (< 20°), the approximate relationship of $1.7^{\Delta} = 1°$ holds true. However, in excess of 20°, this relationship quickly breaks down. Fortunately, when measuring strabismus we are usually dealing with less than 45^{Δ}D, which corresponds to about 25°. Therefore, even up to this deviation, the relationship can be considered to be nearly linear. Many strabismologists simplify this relationship even further, by taking the approximation of 2^{Δ} being equal to 1°.

Index of Refraction

As mentioned above, light travels through a vacuum at a fixed velocity, *c*. It travels through any other medium at a slower speed. The ratio of c divided by the velocity of light through a given medium establishes the *index of refraction* of that medium. Since c (the speed of light) = $\lambda \cdot \upsilon$ (wavelength x frequency), it is apparent that either the wavelength or frequency must also change as light travels through a medium. In fact, it is the wavelength that one is really defining in terms of index of refraction. The frequency does not change.

Certain indices of refraction, which should be kept in mind because they come up often, are as follow:

Vacuum • 1.0 (by definition)
Air • 1.0 (by convention)
Water, Aqueous Humor • 1.33
Cornea • 1.38
Crystalline Lens • 1.42
Vitreous • 1.38
Crown Glass • 1.52
Plastic • 1.49

In other words, since the index of refraction of crown glass is higher than that of plastic, it means light travels through crown glass more slowly than in plastic, or that the wavelength becomes shortened more by crown glass than by plastic. Since the power of refraction is intimately related to the index of refraction (as will be discussed in the next section), crown glass is a more effective lens than is plastic. Accordingly, a plastic lens must be made thicker to produce the same amount of refraction as a glass lens.

Snell's Law

The concept of light being refracted as it passes from a medium with one index of refraction to a medium with a different index of refraction has already been discussed above. However, going beyond this intuitive approach, the mathematical relationship that describes this refraction is the basic law of Snell: $N \cdot \sin i = N' \cdot \sin i'$.

Snell's law describes the behavior of a ray of light which strikes the interface between two media that possess different indices of refraction, N and N'. i and i' are the angles between the ray of light and the "normal", which is defined as perpendicular to the interface. Simple inspection reveals that the greater the index of refraction, the smaller the angle i. However, if one works out the trigonometric function of *sin i* it would equal:

3rd order optics (3 major aberrations) - spherical, chromatic, coma

↓

$$\sin i = i - \frac{i^3}{3!} + \frac{i^5}{5!} - \frac{i^7}{7!} + \frac{i^9}{9!}$$

↑

1st order optics (paraxial)

Since this is a series that goes on *ad infinitum,* we are required, for simplicity's sake, to cut it off. If we only considers the first component of the series, i, we are describing paraxial, or "first order" optics. This is the simplest approach to refraction, and can be stated as ni = n' i'. However, we can take this one step further and go on to "third order" optics (there is no such thing as "second order" optics). In this approximation, $\sin i = i - i^3/3!$. The second component of this approximation contains in it the three major aberrations (spherical, chromatic, and coma). Rarely would an ophthalmologist consider anything beyond third order optics; such is left to computers and those using more subtle instruments of refraction.

An example of Snell's law can be seen in Figure 33 below. Light passes from air (with an index of refraction $N = 1$) into glass ($N' = 1.5$). The angle i between the light and the normal to the interface is reduced to *i'* by the relationship of Snell's law. Note that *i'* is also measured to the normal.

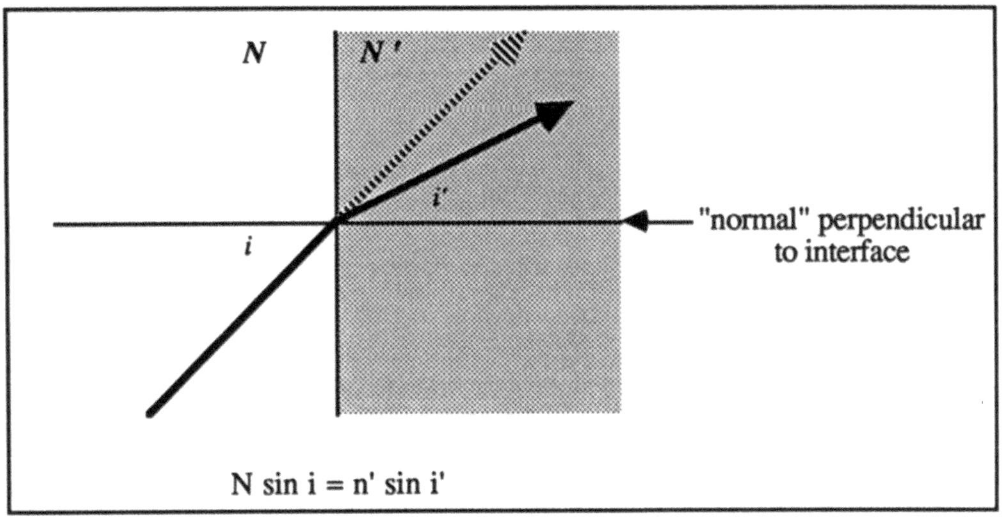

Figure 33 Light passing from a medium of low index of refraction (N) to one with a higher index of refraction (N') is deviated towards the normal line perpendicular to the interface between the media. The amount of deviation is a function of the ratio of the indices of refraction (N'/N).

Critical Angle

As demonstrated in Figure 33, the angle to the normal is reduced when moving from a medium with a low index of refraction to a medium with a high index of refraction. As one would intuitively expect, when light travels from a medium with a high index of refraction to a medium with a lower index of refraction, the angle to the normal increases as in the figure (A) below. There comes a point (B) when the angle to the normal in the media with the higher index of refraction is large enough that it produces an i' of 90°. This is called the critical angle and is, of course, a function of the difference between the two indices of refraction. Beyond this critical angle (C) light is reflected internally. In other words, light never leaves the medium with a higher index of refraction but is reflected back into it.

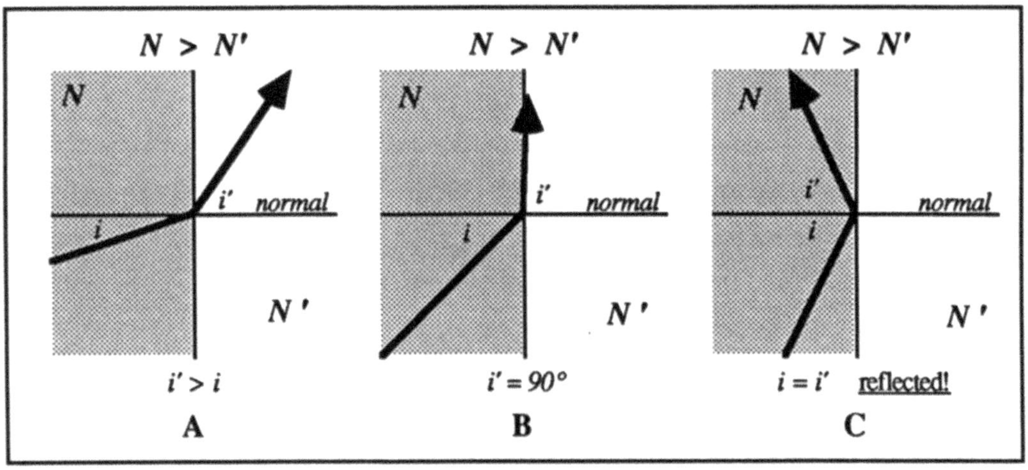

Figure 34 In a situation where light travels from a medium of high refractive index (N) to one of lower refractive index (N'), there is an angle beyond which all light is totally internally reflected and does not escape the first medium — the *critical angle*.

The critical angle, of course, is a function of the difference of indices of refraction between any two media. Between air and water it is 48°. Thus, if one were under water and looked at an angle of 47° to the surface, one could see objects above the water level. Looking at the surface of the water at 48° would produce only a white flash. Looking at the surface of water at an angle of 50° would cause internal reflection and only objects under water could be visualized.

This concept of critical angle has one area of particular clinical importance, in the technique of gonioscopy. As seen below, light emanating from the anterior chamber angle will be totally internally reflected as it strikes the cornea. This is because the difference between the index of refraction of the aqueous (1.33) and that of air (1.0) results in a critical angle of 48°; the geometry of the anterior chamber is such that light emanating fr[illegible] the chamber always strikes the cornea at greater than 48°.

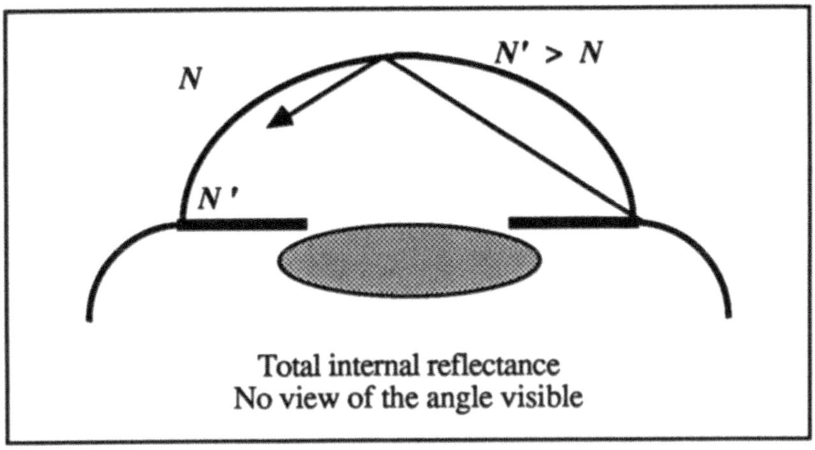

Figure 35 The concept of *total internal reflectance* explains why one cannot view the anterior chamber angle directly.

Two strategies can be used to perform gonioscopy and visualize the angle. In the first strategy, a Goldmann lens, whose index of refraction is approximately the same as that of the cornea and aqueous, is placed on the cornea. Light from the chamber angle therefore travels essentially unrefracted into the Goldmann lens. Here it strikes a mirror (thus the term *indirect* gonioscopy) and is reflected upwards to be viewed through the slit lamp.

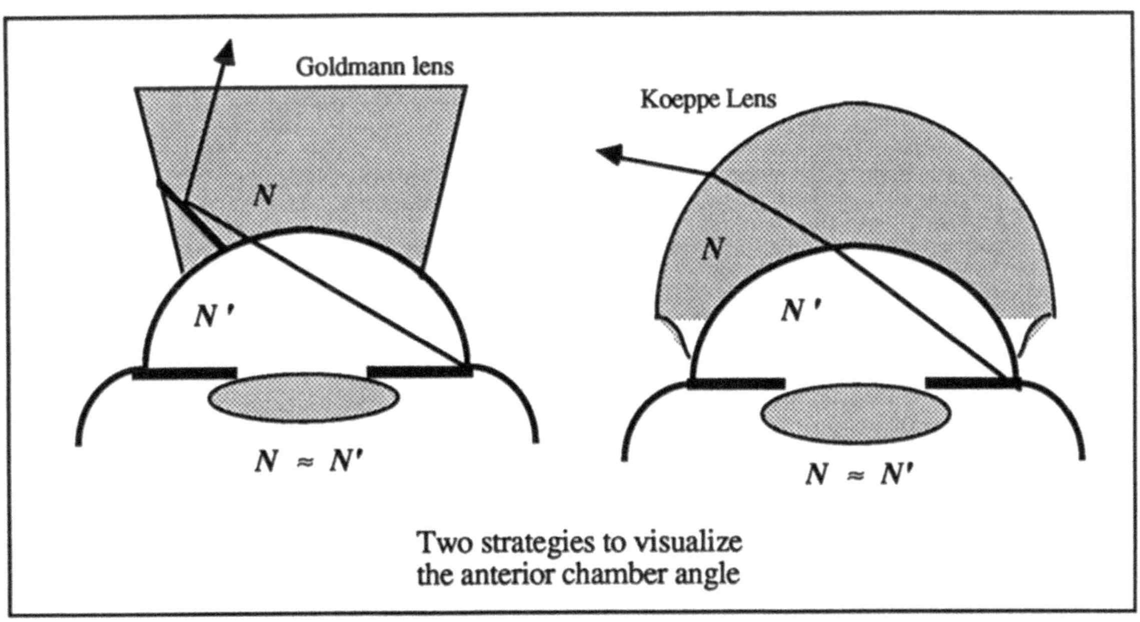

Figure 36 The Goldmann and Koeppe lenses use two different strategies to overcome total internal reflectance.

In the second strategy, a Koeppe lens is used. Again, the key is to have the index of refraction of the Koeppe lens similar to that of the cornea and aqueous. Light travels relatively unrefracted from aqueous through cornea into the Koeppe lens. When it strikes the curved surface of the Koeppe lens it does so at an angle less than 48° and thus, though slightly refracted into the air, can still be seen *directly* with binocular microscopy.

The Power of the Sphere

The power of the sphere is given by the relationship

$$P_s = \frac{N' - N}{R}$$

The value of R is given in meters. Thus, the power of half a glass ball with an index of refraction of 1.50 in air and having a radius of curvature of 25 centimeters would be calculated as in figure 55 to be a power of +2.

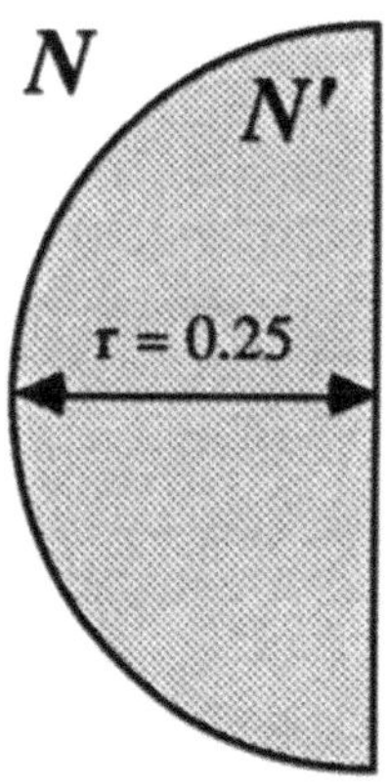

$N = 1.00$
$N' = 1.50$

$$P_s = \frac{N' - N}{r} = \frac{1.5 - 1}{.25} = \frac{0.5}{.25} = +2$$

Figure 37 The power of the sphere (P_S) is derived from the relationship $P_S = (N' - N)/r$.

However, consider the typical biconvex lens in the figure below. Each half of the lens has a radius of curvature of 0.1 and an index of refraction of 1.50. Intuitively, one can see that the power of this lens is going to be two times the power of each half of the lens. Working this out mathematically, consider that in this situation light first travels from air into glass and therefore the relation of P_s equals (N' - N)/1 is computed as (1.5 - 1)/1, giving us a value of +5.

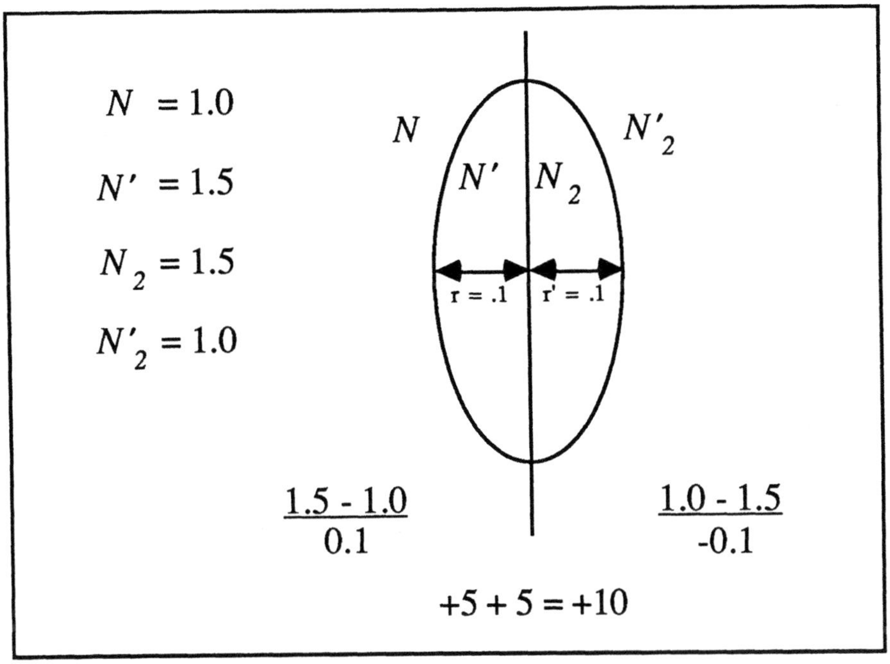

Figure 38 By calculating the power of the sphere for each half of a lens, one can derive the power of the whole lens.

However, light now travels through the second half of the lens into air. The index of refraction of the glass is N_2; the index of refraction of air is N_2'. This time, the radius is in the *opposite* direction. By convention, the radius of a concave surface (as seen by approaching light) has a minus value. Therefore, the power of the second half of this biconvex lens would be (1-1.5)/1 = +5.

Adding the powers of the two hemispheres (+5 +5 = +10) gives us the total dioptric power of the lens.

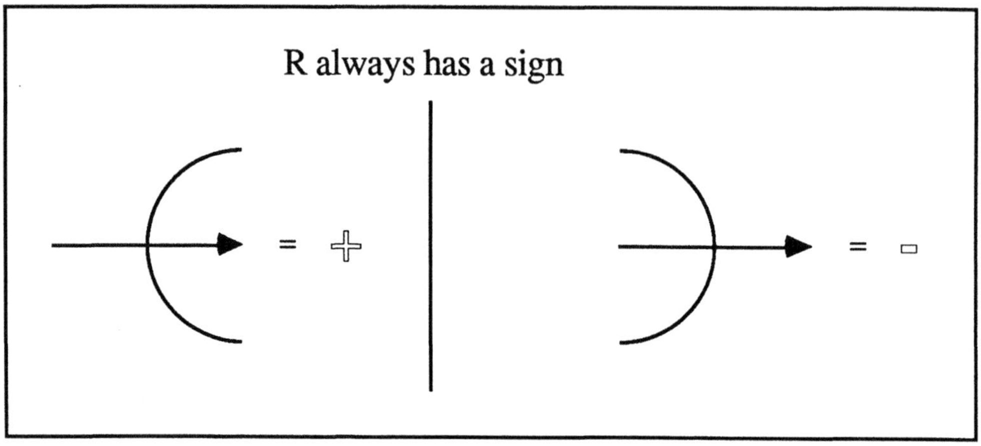

Figure 39 R always has a sign; it is positive (+) for convex surfaces and negative (-) for concave.

It should be emphasized that R always has a sign. As shown above, this sign is positive when light approaches a convex surface and negative when light approaches a concave surface.

The Power of a Mirror

The power of a mirror is represented by the very simple relationship

$$P_m = \frac{-2}{R}$$

Obviously, since a mirror reflects light, the index of refraction is not germane. Consider, as in the figure below, a convex mirror with a radius of 20 centimeters. Because of its convexity, the sign of R must be +. The power of the mirror must be -2/+0.2 = -10.

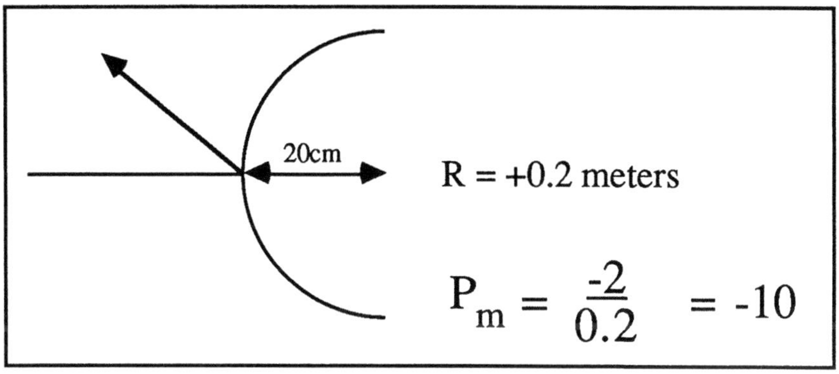

Figure 40 The power of a mirror (P_m) is given by the simple relationship $P_m = -2/r$, where r is the radius of curvature of the mirror.

This mirror will add <u>di</u>vergence to the light that strikes it and reflects off of it.

Problem Set 2

1. What power contact lens do you give a myope wearing glasses with a vertex distance of 15mm if his spectacle Rx were:

    a) -3 D

    b) -6 D

    c) -9 D

    d) -12 D

2. An aphakic patient wears +12.00 D spectacles with a vertex distance of 1.33cm. What power contact lenses would you prescribe?

3. An aphakic patient was given a contact lens of power +12.5, which gave him vision of 20/20, but he can't tolerate the lens. The only spectacles you have to give him are:

    a) +11.5 D

    b) +10.0 D

    At what vertex distance should you set them?

4. An old photographer with a current Rx of +5.00 D spectacles complains of not seeing details clearly in the darkroom under red light. Why? What would you prescribe to correct this?

5. Describe the image:

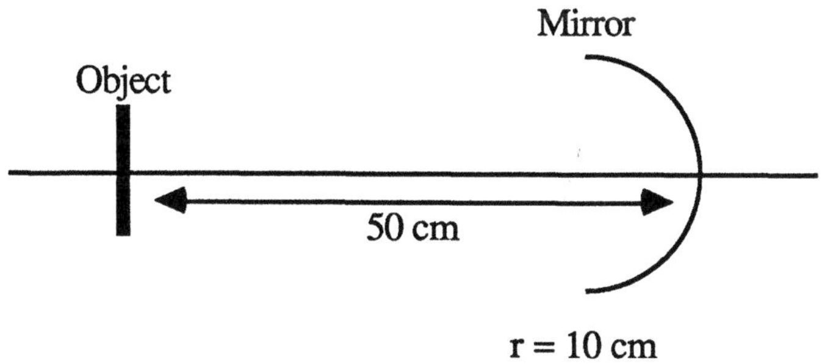

6. Describe the image of yourself reflected off the cornea of your friend, who is standing 0.5 meters away.

7. An emmetrope swimming underwater without the benefit of goggles confronts a shark. What is his refractive error?

8.

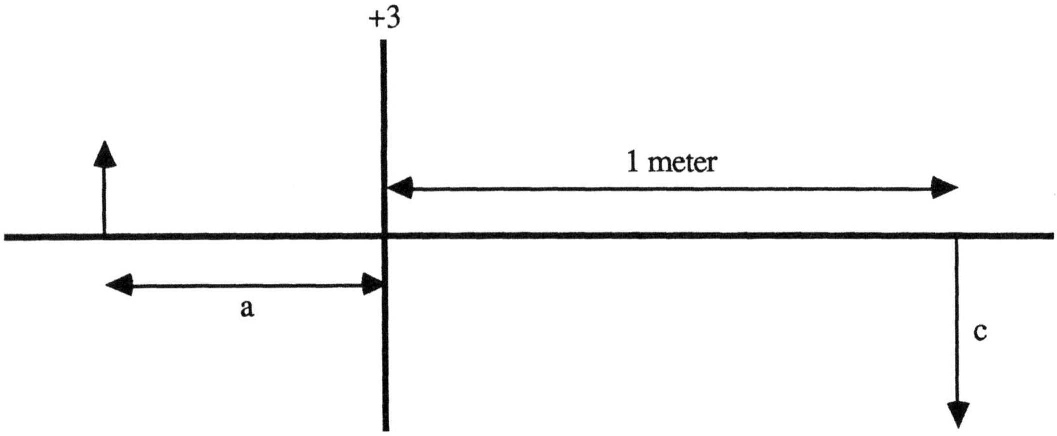

a) What is distance a?

b) Describe the magnification.

c) Draw 3 rays to c.

9. Find the image:

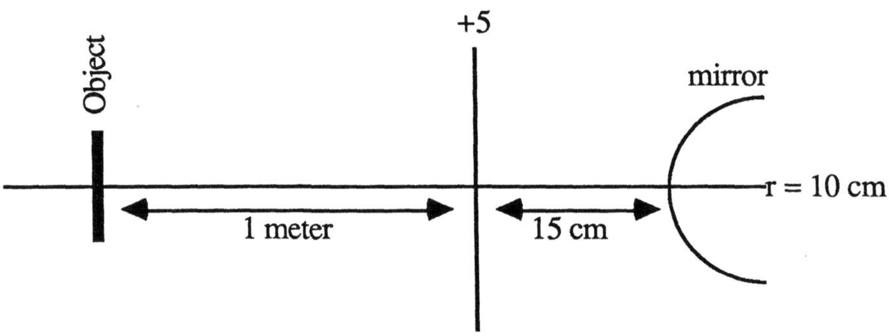

10. A fish quickly evolves from his marine habitat and asks you for an Rx. He is emmetropic underwater. His cornea has an index of refraction of 1.40, and a radius of curvature of 0.7cm. What glasses do you prescribe at a vertex distance of 12mm?

Problem Set 2 — Answers

1.

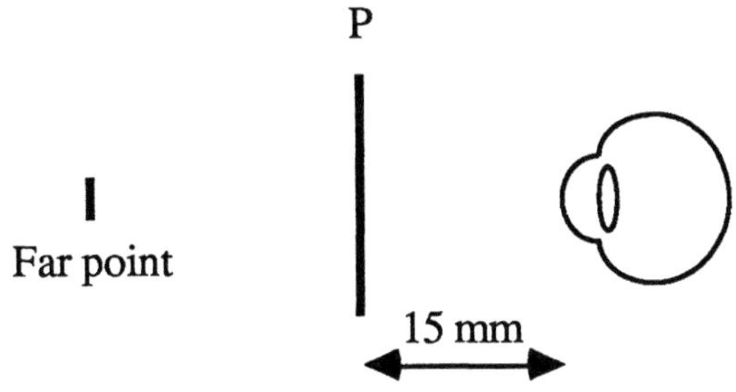

a. If P = -3.00, far point = 33 cm
   33 + 1.5 = 34.5 cm
   1/34.5 = **- 2.9 D** for contact lens
b. If P = -6.00 D, far point = 16.66 cm
   16.66 + 1.5 = 18.76 cm
   1/18.76 = **- 5.5 D** for contact lens
c. If P = - 9.00 D, far point = 11.1 cm
   11.1 + 1.5 = 12.6 cm
   1/12.6 = **- 8.00 D** for contact lens
d. If P = - 12.00 D, far point = 8.3 cm
   8.3 + 1.5 = 9.8 cm
   1/9.8 = **-10.2 D** for contact lens

2.

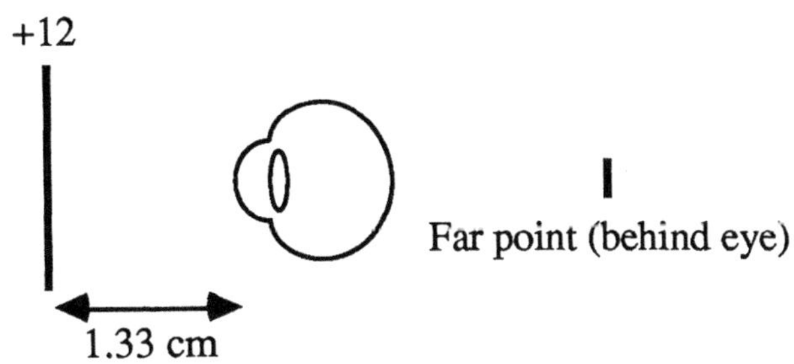

The far point is 8.33 cm **behind** the glasses
8.33 - 1.33 = 7 centimeters behind the cornea

1/7 = **+14.3 D** for a contact lens

3.

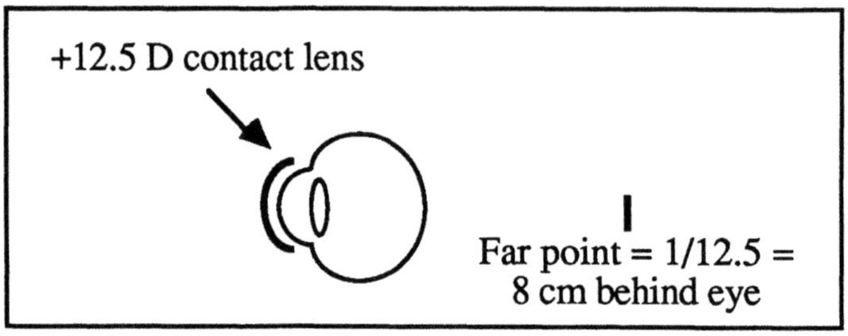

a. +11.5   f = 1/11.5 = 8.7 cm   put the lens 0.7 cm in front (vertex distance)

b. +10   f = 1/10 = 10 cm   put the lens 2.0 cm in front (vertex distance)

4. He is a hyperope; when properly corrected in white light, green (mid-spectrum) light falls on his retina.

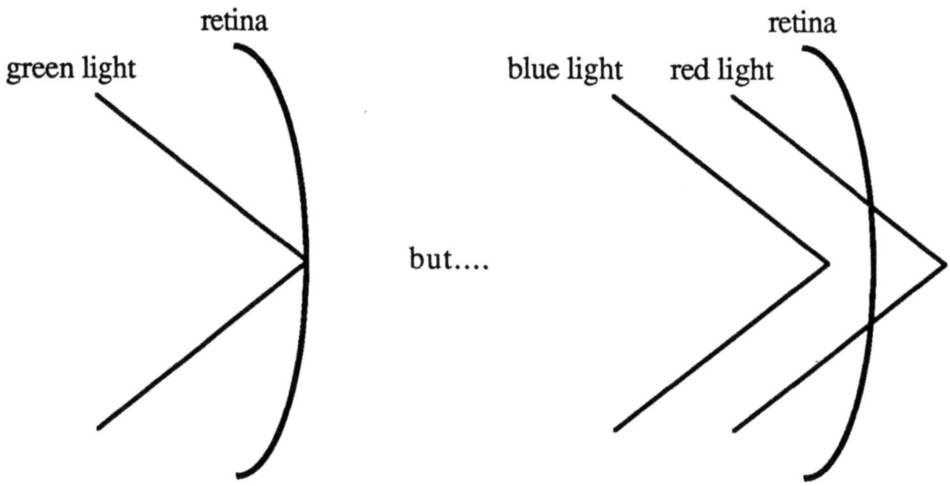

Under red light, he is even more of a hyperope, and unable to focus on red. Give him an extra +0.75 D in his refraction.

5.

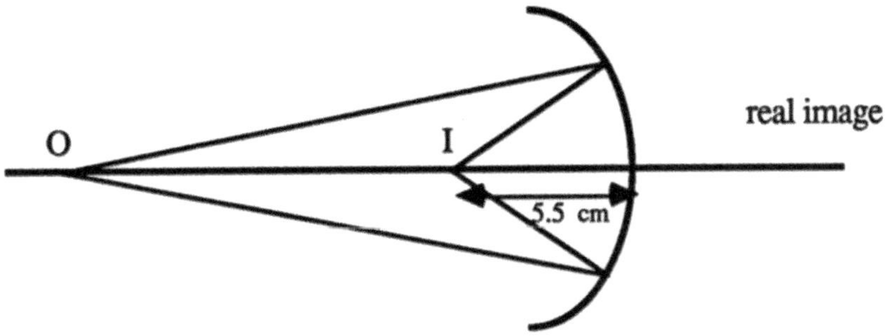

$P_m = -2/r = -2/-0.1 = \mathbf{+20}$

$U + P = V \Rightarrow -2 + 20 = 18$

$1/18 = 0.055$

The image is real and 5.5 centimeters to the left of the mirror.

6. The radius of curvature of the cornea is about 8 mm (or 0.008 meters)
The power of a sphere (P_s) = (N' - N)/r = (1.33 - 1)/0.008 = +42 D
but wait a minute! We want to describe the **reflected** image!

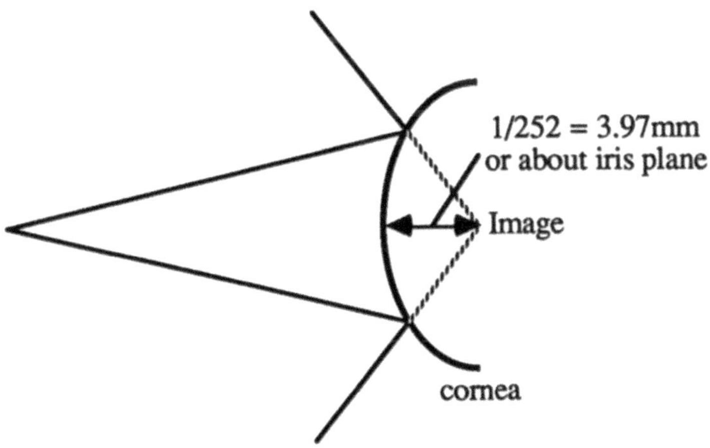

The power of a mirror (P_m) = - 2/r = - 2/0.008 = - 250

$U + P = V \Rightarrow -2 - 250 = -252$

$M = U/V \Rightarrow -2/252 = -1/126$

Thus the image is an inverted, virtual image that appears to "float" about 4 mm behind the cornea. It is 1/126 the size of reality.

7. As we showed in the first part of the problem above, the cornea refracts + 42 D in air. In water, however, N' - N ⇒ 1.33 - 1.33 = **0**! Thus in water, the cornea does not refract at all. This makes the underwater swimmer a 42 diopter hyperope. Even with the +12 D he can crank in by accommodating (he's only 21), he is still left with a 30 D refractive error.

8. a. U + P = -1
   b. M = U/V = -2/1 = -2
   c.

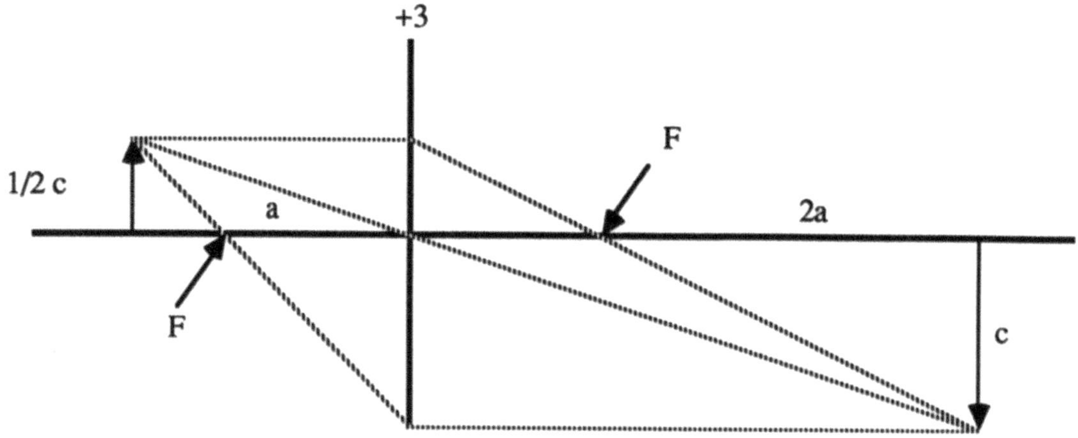

9. The power of a mirror (P_m) = - 2/r = - 2/0.1 = **- 20**

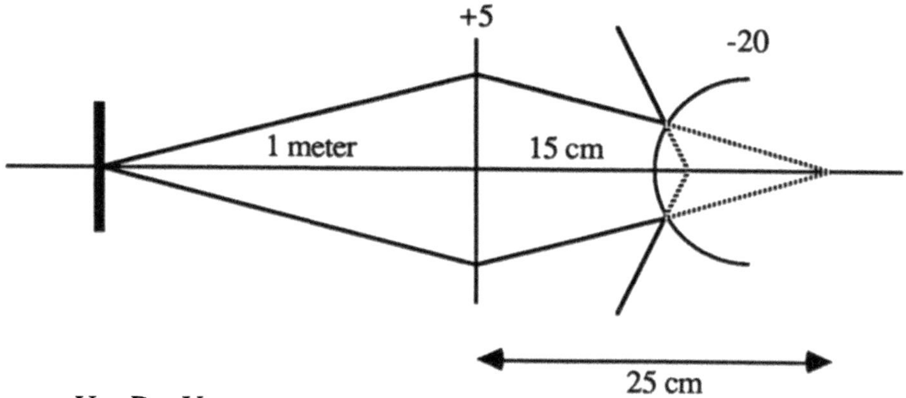

U + P = V
-1 + 5 = +4   1/4 = 25 cm
+10 - 20 = -10

The image is a virtual image, 10 cm to the right of the mirror

10. In water, the fish's cornea refracts:

$(N' - N)/r \Rightarrow (1.40 - 1.33)/0.007 \Rightarrow 0.07/0.007 \Rightarrow +10$ D

In air, his cornea refracts:

$(N' - N)/r \Rightarrow (1.40 - 1.00)/0.007 \Rightarrow 0.40/0.007 \Rightarrow +57$ D

or 47 D more than in water, making the fish 47 D **myopic**!

The far point = $1/47 \Rightarrow 2.1$ cm in front of the eye

$2.1 - 1.2 = 0.9$ cm

$1/0.009 =$ **-111 D** is the power of the spectacles, with a vertex distance of 12 mm, needed to make parallel light come from the far point, thus correcting the fish's new refractive error.

Chapter III — Applications

Refraction

As noted at the end of Chapter 2, refraction can be described by Snell's Law, which states that Nsin i = N'sin i' where N is the index of refraction of the media through which light travels and i is the angle between the ray of light and the "normal", which is perpendicular to the media interface. In first order optics, we can approximate sin i as equal to i simplifying the equation to Ni = N'i'. The power of a sphere (P_s) can be calculated as

$$P_s = \frac{N' - N}{R}$$

For this equation, N is the index of refraction of the medium light travels from; it then enters a second medium (N') which has a curved surface described by it radius (R) in meters.

The power of a mirror is given by the relationship

$$P_m = \frac{-2}{R}$$

Note that the curvature of a refracting or reflecting surface always has a sign. When light encounters a convex surface, R is positive; encountering a concave surface makes R negative. Thus, for the mirror seen in the figure below, the power of R is positive but the power of the mirror (-2/R) is **negative**; light reflected from its surface **diverges**.

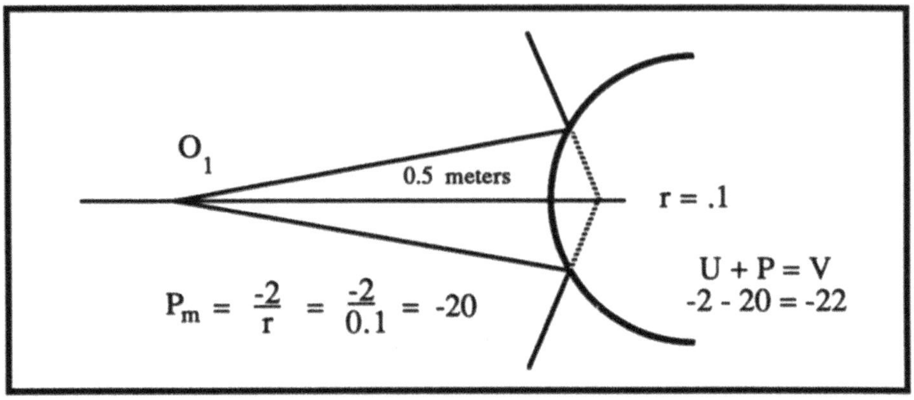

Figure 41 Light from the object point O_1 diverges -2 before striking the mirror. The power of the mirror is -20, imparting further divergence. The virtual image is 4.5 cm (1/22) behind the mirror.

Prism Diopters

The prism changes the direction of the light ray. The angle between the original line of direction and the deviated line is measured in prism diopters ($^{\Delta}$). One prism diopter represents a deviation of 1 centimeter at a distance of 1 meter. In the figure below, light is deviated when it strikes a 10^{Δ} prism. At a distance of 1 meter beyond the prism, the deviation is 7 centimeters below the normal.

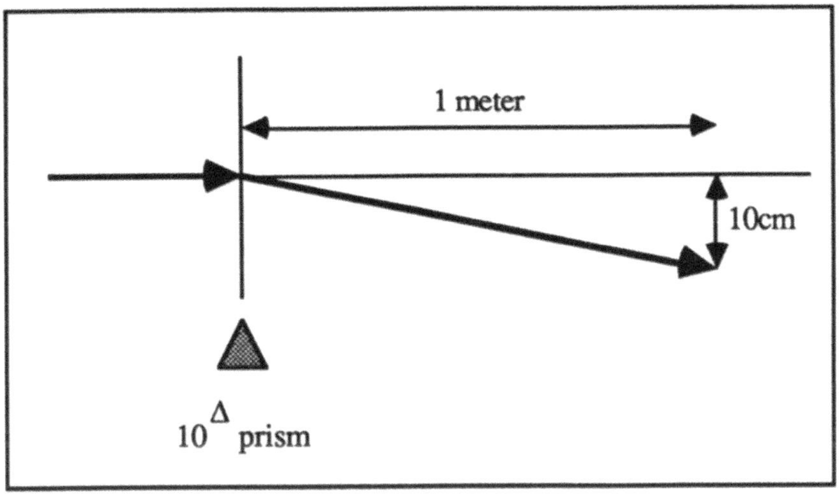

Figure 42 A 10 diopter prism deviates light in the direction of its base 10 cm for every meter the light travels.

Strabismologists are often concerned about determining the ratio between accommodation and accommodation-convergence. They described this as the Ac/A ratio, where accommodation (Ac) is given in diopters (D) and the angle of convergence between the two eyes (A) is given in prism diopters (Δ).

The overall *size* or *thickness* of a prism plays no role in its power, only the angular relationships of its surfaces. A *Fresnel* prism consists of a flat surface onto which have been ground many small, contiguous prisms (see below).

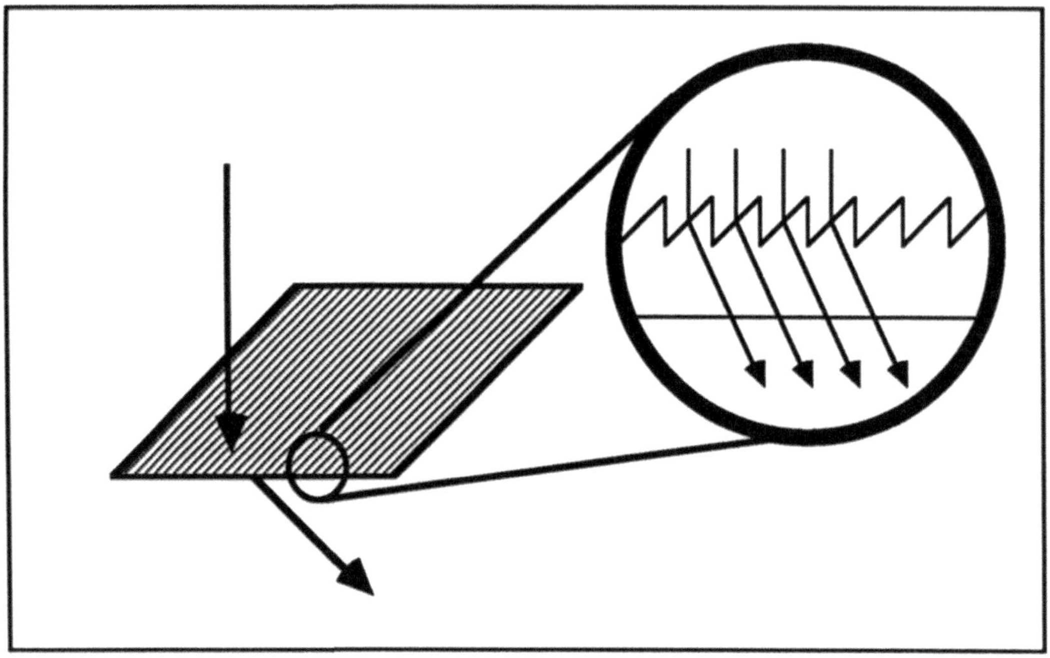

Figure 43 A Fresnel prism consists of rows of prism segments of equal power.

A fresnel prism can be as thin as a moulded sheet of plastic (for example, Stick-On™ prisms used as temporary prisms for spectacles), yet can have significant prism dioptric power. Fresnel prisms, arranged in a concentric, circular fashion, can act as lenses (the stick-on, wide angle lenses placed on windows are examples of this). Whether the bases of the many small prisms are oriented towards or away from the center determines whether the fresnel "lens" has either plus or minus power. This lends intuitive support to the notion, described in Chapter 1, that lenses can be visualized as simply a multitude of stacked prisms.

Prentice's Rule

As noted earlier, a convex spherical lens can be considered as a stack of prisms, each with its base oriented towards the center of the lens. The power of each prism, however, varies with this distance to the center of the lens; the most powerful prism is that at the top or bottom of the lens. Thus it is not surprising that whenever one looks through anything but the center of the lens, in addition to its refractive properties, the lens imparts a prismatic effect as well. When one looks through the visual axis of the lens, however, there is no prismatic effect at all.

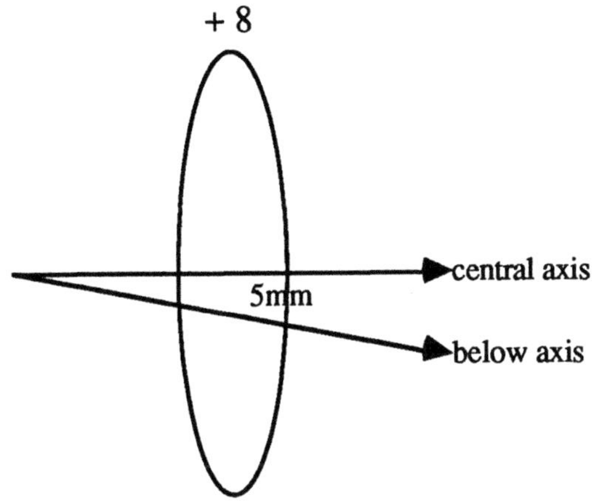

Prentice's Rule

diopters x Distance off axis (cm) = prism diopters

$8D \times 0.5cm = 4^{\Delta}$(base up)

Figure 44 Prentice's Rule states that the extent of prismatic effect induced by a lens is a function of the lens power and the distance off axis of the component involved.

The computation of this prismatic effect follows Prentice's Rule. This simple rule states that this prismatic effect (in prism diopters) is found by multiplying the diopter power of the lens by the distance from the central axis of the lens, given in centimeters. Thus in Figure 44 above, if the lens is a +8 lens and the patient is looking half a centimeter below the axis of the lens, $0.5cm \times 8D = 4^{\Delta}$; there exists a 4^{Δ} prismatic effect. Note that since the deviation is below the central axis and the prismatic effect of a plus lens is always with the base of the prism towards the central axis, that this patient is going to be subject to 4^{Δ} **base up** by looking as he does.

Similarly, prismatic effects can occur when the patient looks through a minus lens. Moreover, this same prismatic effect occurs when the patient looks above the central axis or to either side of the central axis. Again, Prentice's Rule describes the amount of prism such an effect produces.

One situation in which this rule is of great importance is in the use of bifocal segments. It is obvious that a patient will develop more and more prismatic effect as he looks further and further below his visual axis. At a certain point he will encounter the bifocal segment. The prismatic effect of the bifocal segment is going to be directly related to the distance between where the patient looks through the bifocal segment and its central axis. Consider Figure 45 below. A person with -6.0 spherical spectacles has a reading add of +2.50D in the inferior portion of the spectacles. When he looks through the central axis of the spectacles, there is no prismatic effect. As he begins to look below the central axis, there is a prismatic effect with the base down. Just before getting to the add segment of the bifocal spectacles, the prismatic effect would be .6 x 6.0 = 3.6$^{\Delta}$ base down. However, when the patient looks 1mm below this line he looks through the add component of their bifocals 6mm from the central axis of the +2.50 component. This gives 2.50 x .6 = an additional prismatic effect of 1.5 diopters base down. Suddenly, the wearer goes from 3.6$^{\Delta}$ to 5.1$^{\Delta}$ of total prism base down. The result is a "jump" of the image upwards.

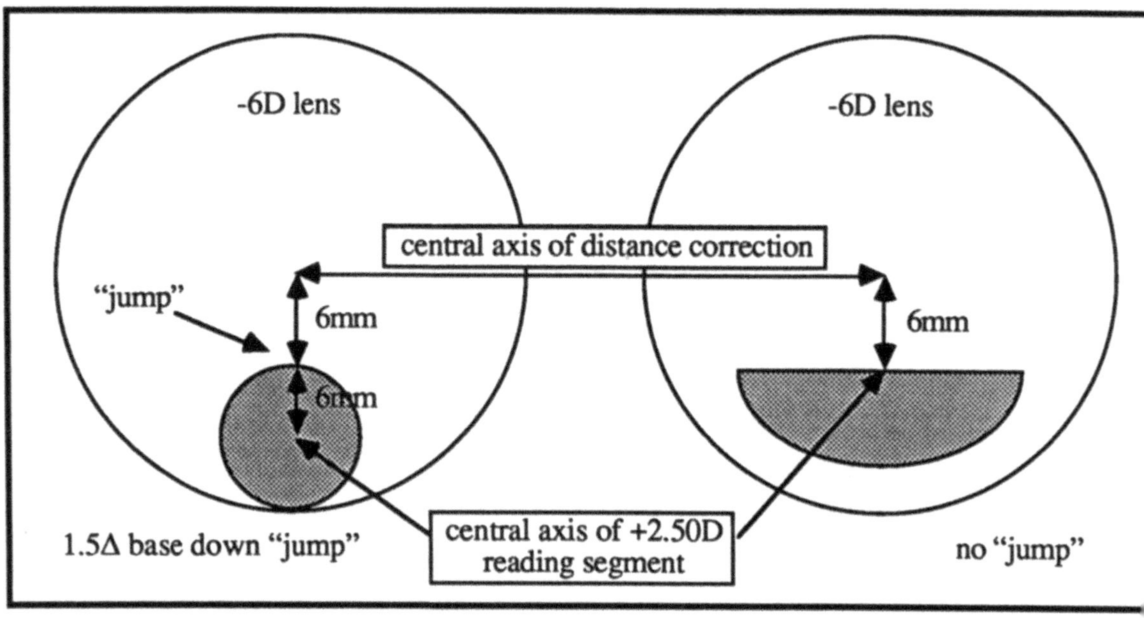

Figure 45 The "jump" in image produced by going from the carrier lens to the bifocal component is minimized by the use of a flat-top reading segment.

One way to avoid this image jump is to use a "flat top" add. In this arrangement, the bifocal add is created so that the visual axis is at the top of the bifocal component. As a patient looks further down from the central visual axis of the lens, there is a greater and greater base down prismatic effect. However, when they get to the +2.50 bifocal add, since they are at their visual axis, there is no additional prismatic effect. In other words, Prentice's Rule for the bifocal would be 2.50 x 0 (the distance from central axis) = 0. This is the major advantage of having a flat top type of bifocal segment.

Cylinders

Until now we have only considered the power of a spherical surface with a single radius of curvature. It is obvious that because of the radial symmetry of a sphere, light striking any part of the surface will come to the same focal point. However, a refracting element can be irregular. One relatively simple type of irregular refracting element is called a *toroid*. A toroidal surface has two fixed radii of different magnitudes — a common example is the surface of an egg.

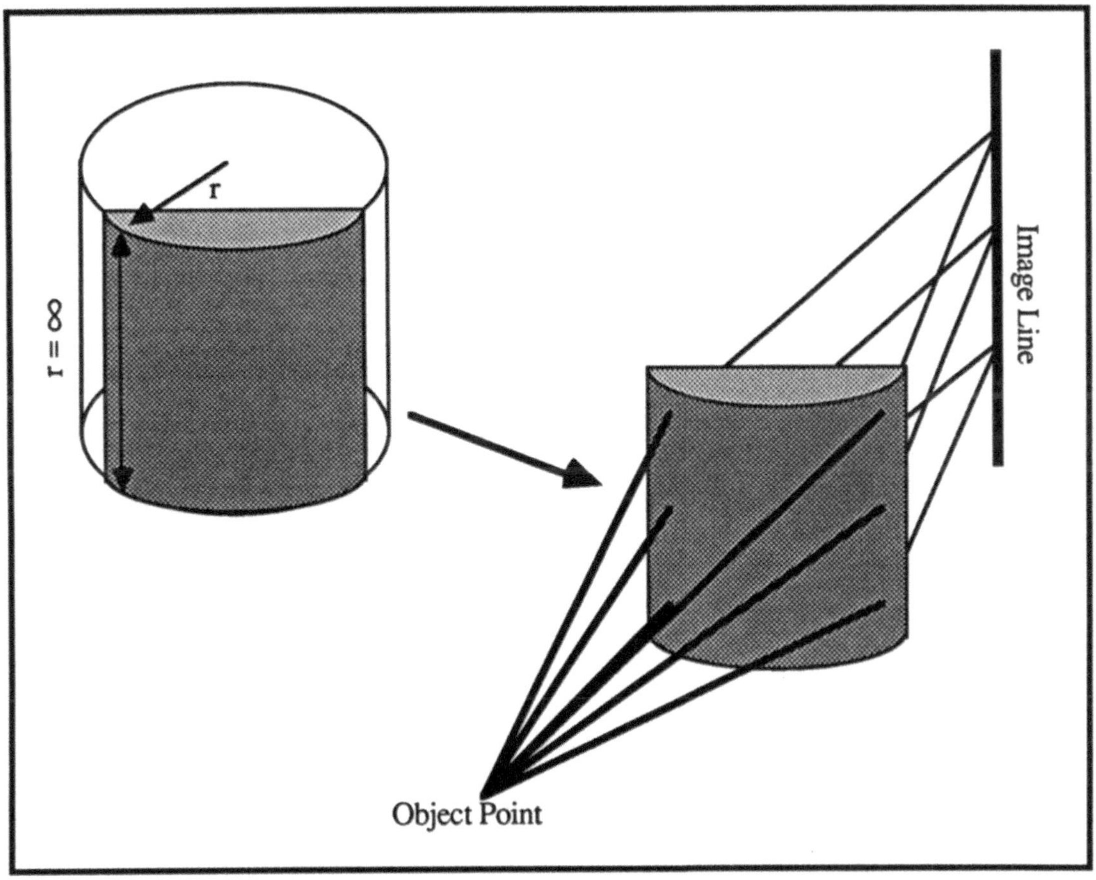

Figure 46 A section through a cylinder, parallel to the cylinder's axis, produces a curved surface with curvature in only one axis. Such a cylinder refracts light from an object point to an image **line** parallel to the cylinder's axis.

The simplest toroidal surface, shown in Figure 46 above, is a cylinder cut in longitudinal section. Note that a radius of curvature is presented only in one plane. In the other plane there is no curvature, hence the radius can be considered to be infinite. Recall that the power of a refracting surface is (N' - N)/r. In the case of the cylinder above, the vertical surface has an infinite radius and thus has no refracting power. In the example above, an object point diverges light to a cylinder, which produces convergence only in one axis. There are an infinite number of image points which align themselves along a straight line — the *image line*, which is parallel to the axis of the cylinder. In this example, a real image would be formed along the vertical image line; if the cylinder was in a horizontal position, a horizontal image line would be formed instead.

One concept that often causes confusion among students of optics is that the power of a cylinder is perpendicular to its axis. Consider the cylinder in figure 46 above — its axis is vertical (90°). Yet the refractive power of this cylinder is only going to exist where the radius of the cylinder has a real value, in the horizontal plane. It is precisely because the power is in the horizontal axis that the points remain aligned in a vertical axis — there is no vertical power to bring them completely together.

When a lens is described by an ophthalmologist, it is the **axis** of the cylinder that is usually described, not the power. Thus when we talks about having a prescription of +1.00 + 0.50 x 90°, we mean that there is one diopter of plus power in all axes and, in addition, half a diopter of cylinder at **axis** 90°. However, keep in mind that the **power** of this cylinder is really at 180° (orthogonal to the axis).

The Power Cross and Transposition of Cylinders

In order to keep clear the distinction between the power of a cylinder and the axes, it is useful to be familiar with the notation of the power cross and, when in doubt, revert to it as a means of calculating cylinder. The power cross is usually illustrated only in the axes of 90° and 180°. However, keep in mind that it can represent any two axes orthogonal to each other. Consider the power crosses depicted in Figure 47 below. Power cross *A* represents a spherical lens with the power of +3 in all axes. Power cross *B* represents a *combination* of spherical and cylindrical power, with more plus at axis 180°.

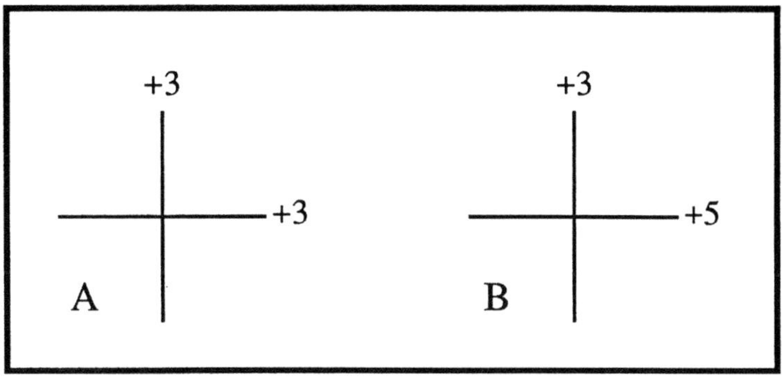

Figure 47 **Power Crosses** — For A, the cross represents a spherical lens with equal power (+3) in every axis. For B, the power at 180° is 2 D more plus than at 90°.

When dealing with a power cross with combined cylindrical and spherical characteristics, manipulations are best accomplished by separating these two components. In Figure 48 below, the power cross of +3 at 90° and +5 at 180° can be considered to be equal to a spherical lens of +3 in all axes added to a cylindrical lens with a power of +2 in the 180° axis.

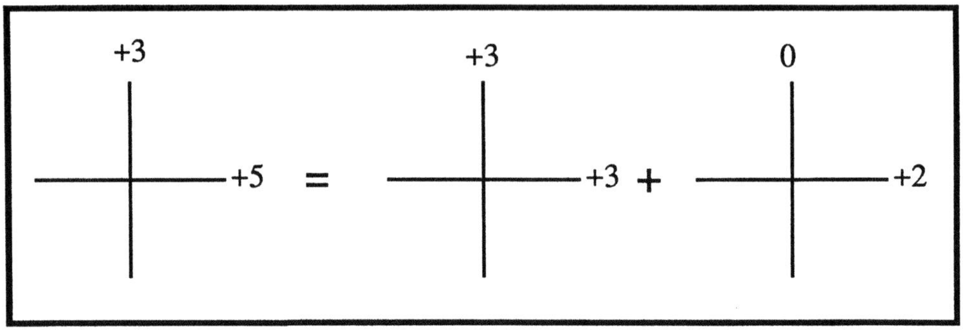

Figure 48 The unequal axis power cross can be broken down into the general (spherical) component plus the additional power in one axis (in this case, +2 at 180°)

The power of a cylinder can be either plus or minus. Thus the combined spherical and cylindrical lens above can be broken down into its two constituent parts in a different way. Using a negative value for the cylindrical power, +3 at 90° and +5 at 180° can be represented as a spherical lens of +5 in all axes added to a cylindrical lens of power -2 at axis 90°, as shown below (Fig. 49).

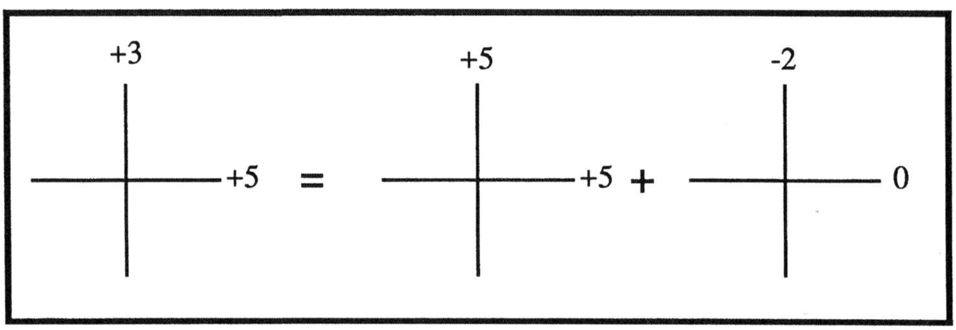

Figure 49 The power cross can be broken down in two ways. Here the inequality between axis powers is given in minus cylinder (-2 at 90°) instead of in plus cylinder, as in Figure 48.

This brings us to the concept of transposition of cylinders. As you know, ophthalmologists, optometrists and opticians can write their prescriptions in either plus or minus cylinder. There are a variety of conceptual and practical advantages to each. In years past, an optician started with a spherical lens blank obtained from an optical supplier and ground into that lens any cylinder component required. In the figure below, a cylindrical lens of -5.00D x 180° is ground into the back surface of a +4.00 D lens blank. Two power crosses can be used to represent the front and back surfaces, and they can be combined algebraically to represent the entire lens.

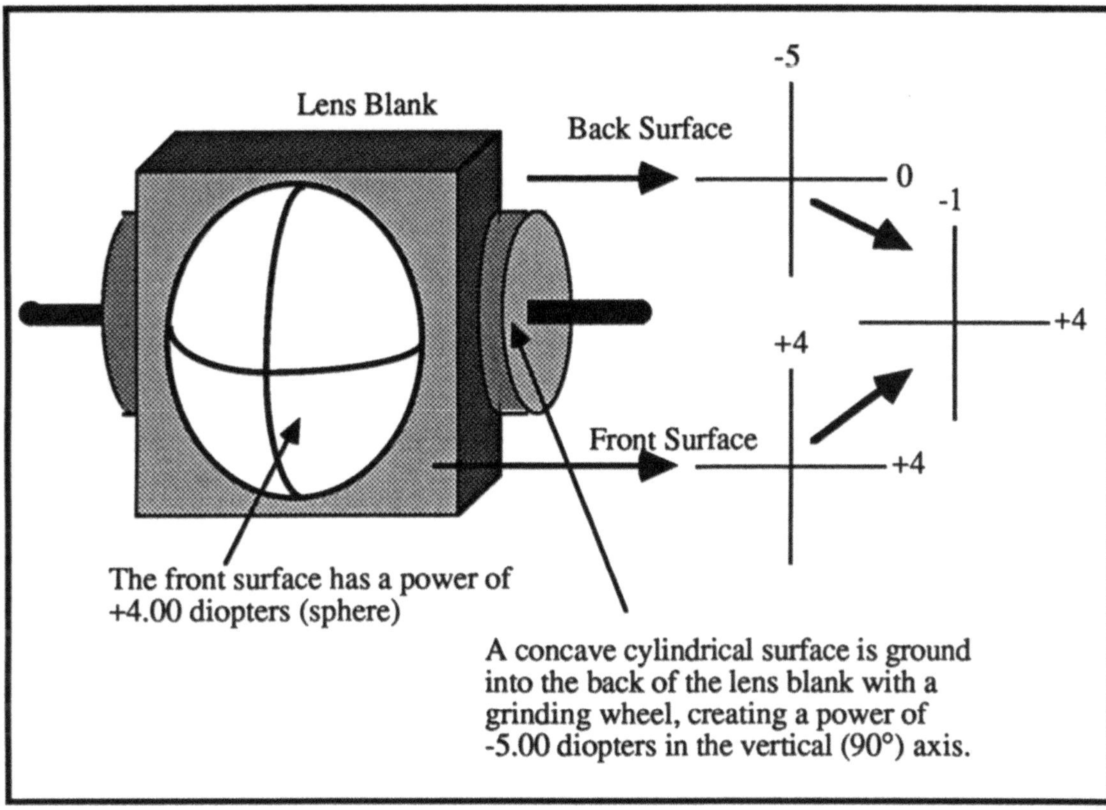

Figure 50 Taking a plano-convex lens blank, the optician grinds a cylindrical convex lens into the plano surface. This gives rise to a sphero-cylindrical lens that can be described by the power crosses above.

Because it was technically easier to grind a concave surface into a lens (creating a *minus* power), "minus cylinder" notation could be used directly by the optician. Today, opticians keep on hand a large variety of sphero-cylindrical lens blanks. By aligning the axis of the blank to that in the prescription and then cutting the blank to fit the frame, a simple prescription can be fabricated without grinding, and so the form of notation has become less important.

Therefore, the choice of notation now usually just comes down to a matter of habit as to what a particular individual or institution uses, or to whether the trial frame or phoropter is limited to plus or minus cylinder lenses. Nonetheless, the competent ophthalmologist must be able to convert with facility from one mode of expression to the other. Consider the refraction +4.00 - 1.00 x 180°. Obviously, this is written in minus cylinder. How might it be transposed into plus cylinder? One can do this either by means of a formula or through the use of the power cross. The latter permits a greater appreciation for the nature of cylinder, but the formula is quicker. It simply states that the new sphere is the sum of the old sphere and cylinder and the new cylinder is the same power as the old cylinder but at an axis 90° to the old cylinder. Thus using the formula, +4.00 -1.00 x 180° is transposed to +3.00 (+4.00 - 1.00) + 1.00 x 90°. However, let's do the same exercise with a power cross as demonstrated below.

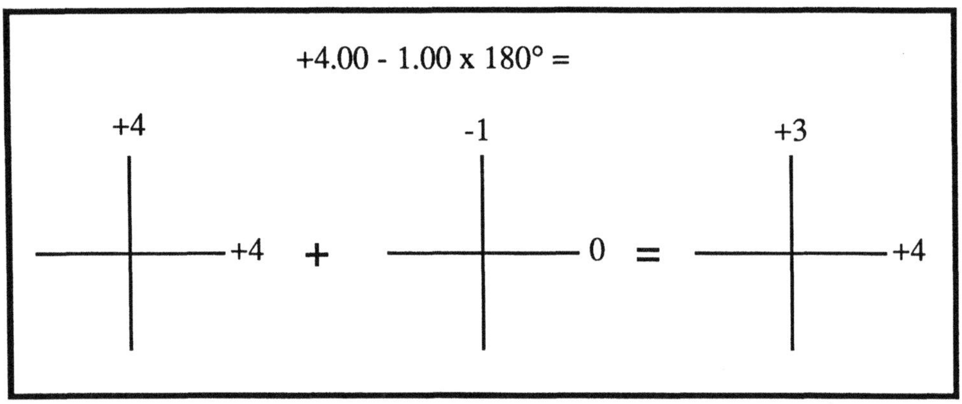

Figure 51 To transpose cylinders using the power cross, first draw out each component of the refraction as power crosses (remember, the **power** of a cylinder is perpendicular to the **axis**); then add them algebraically.

A +4.00 is depicted as a power cross with +4 at axes 90° and 180°; -1.00 x 180° is shown in the power cross as -1 at the 90° axis! Remember, the *power* of a cylinder is 90° from its axis. Therefore, the refraction of +4.00 -1.00 x 180° is seen above to be the combination of the two power crosses shown. However, the algebraic addition of each axis gives us a power cross of +3 at 90° and +4 at 180°.

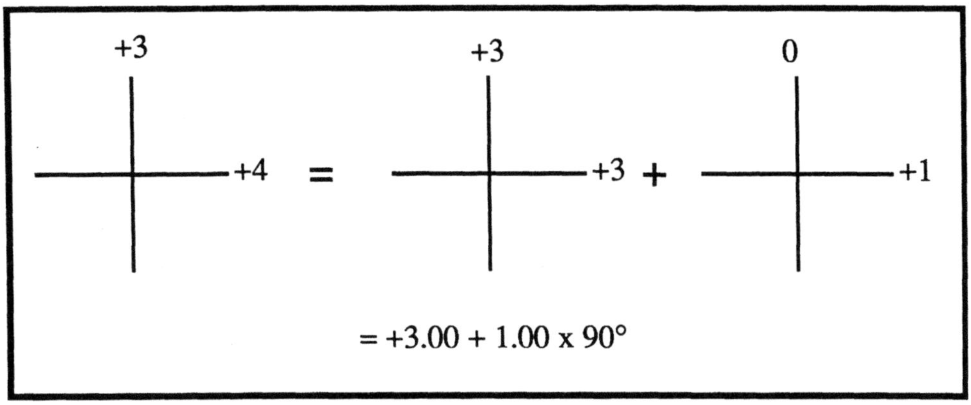

Figure 52 You can break down the power curve into its constituent parts using either plus or minus cylinder.

In the figure (52) above, we can see how +3 at 90° and +4 at 180° can also be divided into two constituent components, spherical and cylindrical. This produces +3 at all axes and +1 at 180°. However, we're still talking about the power of a cylinder. Thus, this is written out as +3.00 + 1.00 x 90°.

Going from a plus cylinder to minus cylinder is equally straightforward. Consider the refraction of -2.00 -0.50 x 30°. By formula, this can be quickly converted to -2.50 + 0.50 x 120°. However, it remains best to conceptualize either of these two refractions as the power cross shown below. Remember that the -0.50 x 30° represents an additional -0.50 at 120°, as shown in the power cross.

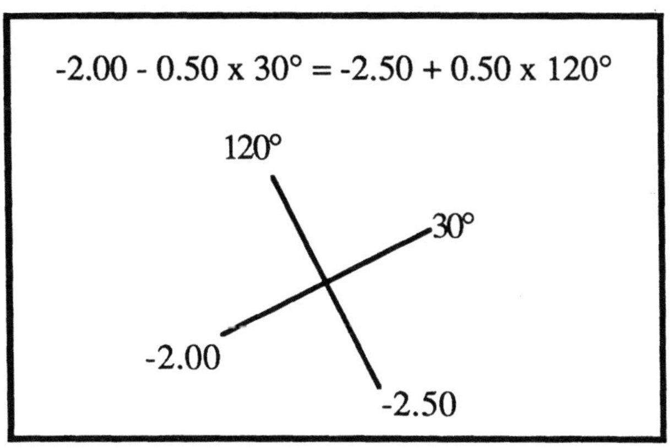

Figure 53 This power cross can be broken down into constituent parts to describe the lens in plus (-2.50 + 0.50 x 120°) or minus (-2.00 - 0.50 x 30°) cylinder.

Magnification

We've already seen how magnification (M) is proportional to the vergence power of an object (U) and inversely proportional to the vergence power of the image (V) for any lens system, giving rise to the relationship M = U/V.

So it is with any image that is refracted. However, a simple form of magnification is obtained whenever an object is approached. Clearly, the angle subtended by a 10 inch object at 10 feet is half that of the same object at 5 feet. One could say that viewing the object at 5 feet provides 2x magnification. It follows that examining the object at 1 foot would provide 10x magnification, etc. Because most individuals scrutinize a small object at a distance of 25 centimeters (one-quarter of a meter), it has come to be convention that the magnification provided by a strong plus lens is described as the diopter power of that lens divided by 4. Hence, a hand held magnifier of 10 diopters provides a magnification of 2.5; the magnifying power of a +20D Nikon indirect lens often used to scrutinize Kodachromes is 5.

Galilean Telescope

A Galilean telescope provides magnification with two lenses of a particular power arranged in a particular position. Although historically this telescope has been used for astronomical purposes, it turns out to be very important to the clinician. Not only are Galilean telescopes used to provide magnification for patients as low vision aids, but more importantly, the Galilean telescope is in fact built into our own ocular system any time a refracted error is corrected. Because of this, the properties of this Galilean telescope are constantly evidenced and need to be understood.

Consider, as in Figure 54 below, two lenses arranged in such a way that light with a vergence of 0 entering the system also exits with a vergence of 0. The light first strikes the +10 lens, giving it a convergence of +10. However, it then encounters the -20 lens after traveling only 5 centimeters. This convergence of +20, when added to the power -20 of the second lens, causes the light to exit with the vergence of 0. As you will note from the figure, however, the light does not leave exactly as it entered the system. The difference is pure magnification. The power of this magnification in the Galilean system is always $-P_2 / P_1$,where P_1 is the power of the first lens that the light encounters. In the example of Figure 54, the power of this Galilean telescope would be +2. Since this number is positive, the image is upright (as it always is in a Galilean telescope) and because the number is greater than 1, there is true magnification.

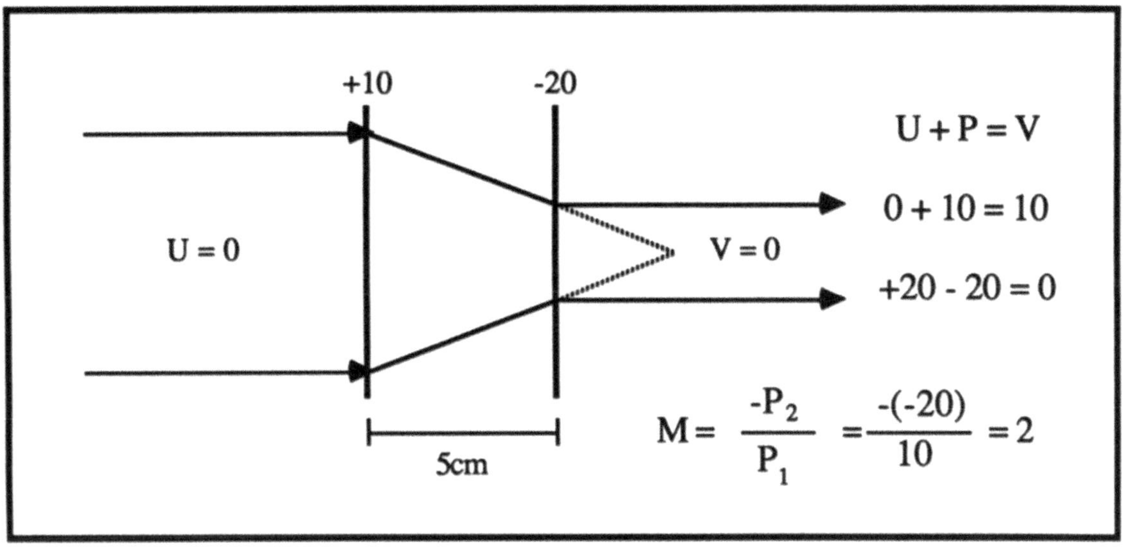

Figure 54 For a Galilean telescope, light entering with a vergence of 0 exits with a vergence of 0. With this arrangement of lenses, the magnification is 2.

One of the clinically important aspects of a Galilean telescope is that the eye itself can be considered to be one of the two lens systems in a Galilean telescope and any optical device used to correct for refractive error to be the other lens. Since the light travels towards the eye, either the spectacle or contact lens would be regarded as the P_1 lens, and the eye itself as the P_2 lens. Consider as in Figure 55 below, the aphakic eye which would be corrected with a +16 diopter intraocular lens. For purposes of a Galilean telescope, this eye is considered to be a -16 diopter lens system (since it is -16 with respect to emmetropia), with the lens being at the nodal point. Such an eye has a far point of 6.25 centimeters to the right of the nodal point. A pair of spectacles 1.7 centimeters in front of the cornea (vertex distance = 1.7 cm) would in fact put the spectacles 1.75 centimeters in front of the nodal point (1.7 + .05), and so the total distance from the glasses to the far point would be 8 centimeters. The appropriate refractive correction at such a vertex distance would be the reciprocal of 8 centimeters or +12.5. This would be the appropriate power lens in a pair of glasses for this aphakic patient. The interesting thing is that this lens system now constitutes a Galilean telescope, with P_1 equaling +12.5 and P_2 equaling -16 . Magnification is $-P_2 / P_1$, which equals 16/12.5 or +1.28. Thus, correcting the aphakic patient in this manner with spectacles produces a magnification of 28%. It is no wonder that with the central magnification being increased by 28% there is by necessity a ring scotoma around the edge of the magnified zone.

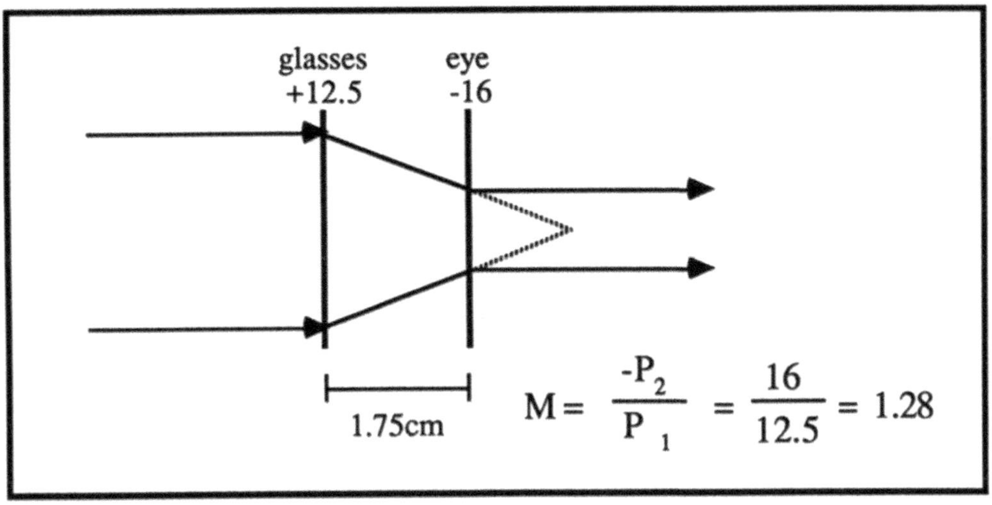

Figure 55 Correction of an aphakic eye with spectacles creates a Galilean telescope. This results in a magnification of +1.28.

It follows, then, that when we deal with changes in vertex distances, we are in fact manipulating the lenses of a Galilean telescope and changing the magnification as well. In general, the refractive error of the eye represents P_2, and the corrective spectacle or contact lens represents P_1.

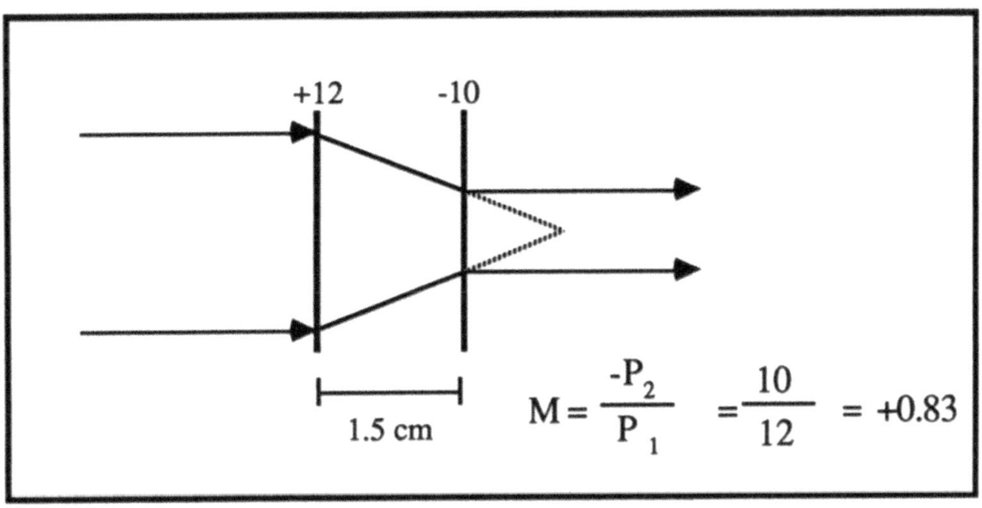

Figure 56 The high myope wearing corrective lenses sees a minified image (in this case 0.83x) due to the Galilean telescope of the combined eye/spectacle lens system.

Let's consider this example in the myopic correction shown in Figure 56 above. A -10 contact lens wearer develops corneal problems and requires spectacle correction. If the spectacles were to have a vertex distance of 1.5 centimeters, the appropriate correction would be (as derived by the vertex distance calculations described above) approximately -12 diopters. The Galilean telescope magnification would be $-P_2 / P_1$, which in this case would be 10/12 or +0.83. The "magnification" of this myope looking through spectacles would be 83% of the original size — in other words, there would be *minification*. This might prove dissatisfying to the patient who not only would be perturbed by seeing things smaller than expected but would actually lose a little visual acuity due to the minification effect.

On the other hand, minification is not always a disadvantage and magnification is not always an advantage. Suppose that this patient wore such a correction until the age of 45. At that time, she comes into a contact lens clinic and wishes to correct her refractive error with a contact lens (which, continuing the example above, would be a -10 diopter contact lens). The fitting is done well and the patient is initially delighted with the fact that she not only has improved her appearance but now sees everything 20% bigger than before which apparently improves visual acuity as well. In addition to improving her cosmesis, the contact lenses give her a bigger, clearer world. But now comes the rub.

The amount of accommodation expended through any Galilean telescope equals the amount of accommodation required without the use of the telescope times the square of the magnification provided by that telescope. Returning to figure 56, we see that the patient wearing spectacle correction for her myopia only needed to accommodate 69% as much as she would have had she been an emmetrope. Moreover, this is 69% as much accommodation as she would have required with contact lens correction (which essentially eliminates the Galilean telescope and its magnification). However, remember that a 45-year-old patient has reduced capacity for accommodation (see Figure 21). This particular patient has previously been able to resist the symptoms of presbyopia quite well, since she has only been required to use 69% as much accommodation as normal, but now that she has converted to contact lenses, she has to put in the remaining 31% accommodation. This extra demand on accommodation is often not successful in this setting. Because of this it is not uncommon that high myopic patients do very poorly when converting from spectacle correction to contact lens correction in middle age — they may see extremely well at distance, but suddenly develop the need for reading glasses at near. Not surprisingly, these patients are not appreciative of these optical effects and attribute the sudden development of presbyopic symptoms to some error in the ophthalmological care that they have received. Beware of giving contact lenses to the myopic middle-aged patient who has worn spectacles for years!

Problem Set 3

1. A flashlight drops into a cylinder of water, with the light coming to rest 33 cm below the surface, pointing up. The water surface has a concave meniscus with a radius of curvature of 50 cm. The index of refraction (N) of water is 1.33. Describe the image.

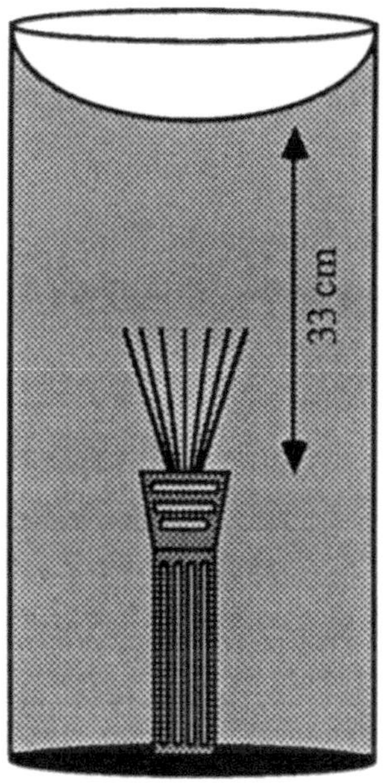

2. Would substituting saline (N = 1.33) for vitreous (N = 1.4) make an eye hyperopic or myopic?

3. What would the image displacement be with a 20^{Δ} at 6 feet?

4. Looking through a +12.00D aphakic spectacle 15mm below the optical center produces what prismatic effect? Does the image appear above or below reality?

5. A -6.00D myope OU also has a 10^{Δ} esotropia. How could you cure his diplopia without prisms or surgery?

6. Transpose the following spectacle prescriptions into minus cylinder:

   a. +3.00 + 1.50 x 90°

   b. -2.50 + 3.00 x 180°

   c. Plano + 1.25 x 45°

   d. 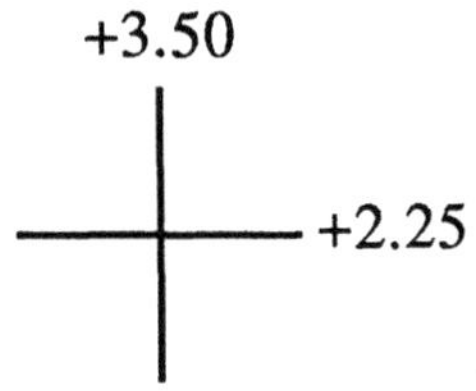

7. A scotoma on a tangent screen suspended 2 meters in front of the patient measures 20 cm across. What is the size of the lesion on the retina?

8. A low-vision patient with 20/200 acuity comes to your office with a Galilean telescope mounted on a spectacle frame; he picked up the device at a rummage sale, and finds it doesn't work. You discover that the objective lens, which in this device must be mounted 5 cm in front of the eyepiece, is missing. The eyepiece has a power of -

Problem Set 3 — Answers

1. *N* (water) = 1.33; *N'* (air) = 1; R = 0.5 and is positive (concave to arriving light)

   The power of the meniscus is: $\frac{N' - N}{r}$ which equals $\frac{-0.33}{0.5}$ or -6.6

   U + P = V ⇒ -3 - 6.6 = -9.6

   M = U/V ⇒ -3/9.6 = 0.31

   The image is virtual and appears 10.5 cm below the water level and is minified to 31% of reality.

2. In substituting saline for vitreous, we change the optical properties at the lens-vitreous interface.

   The power of this interface is: $\frac{N' - N}{r}$

   where *N'* is the vitreous replacement and *N* (lens index of refraction); r remains unchanged. Suppose the radius of curvature of a lens is .01 meter. Then r in this case equals -.01 since the back of the lens curves concave to the direction of light. With vitreous, the back surface of the lens has a power of:

   $$\frac{1.40 - 1.42}{-0.01} = \frac{-.02}{-.01} = +2$$

   Substituting saline, we have:

   $$\frac{1.33 - 1.42}{-0.01} = \frac{-.09}{-.01} = +9$$

   Thus the power of the lens has **increased**, making the patient more **myopic.**

3. 6 feet is equivalent to about 2 meters. Therefore, 2 meters x 20^{Δ} equals 40 cm of displacement.

4. Using Prentice's Rule, 12 D x 1.5 cm = 18^{Δ} ; the prism is base **up**, so the **image** is displaced **downwards!**

5. A myope looks through spectacles which can be schematized as base out prisms whose apices meet at the optical center (see diagram). By decreasing the PD (distance between optical centers) we effectively induce base out prism when looking straight ahead.

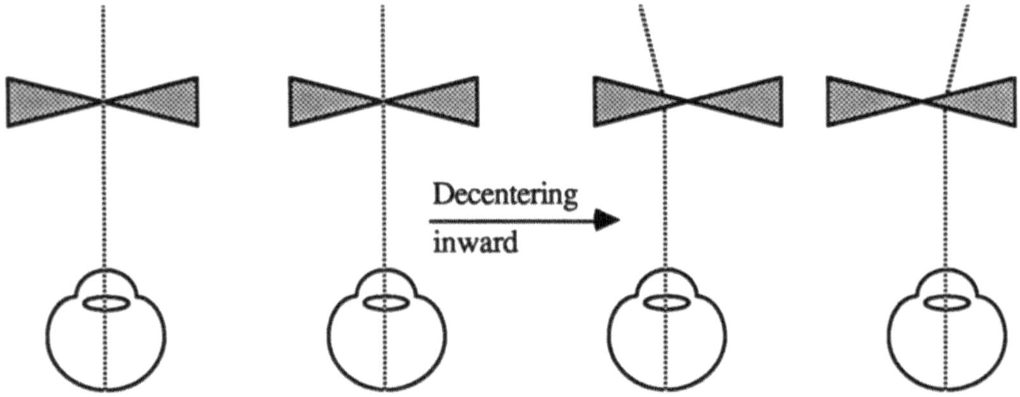

In this patient, we wish to correct an esotropia of 10^{Δ} and so we need to induce 5^{Δ} of base out prism in front of each eye. Using Prentice's Rule,

-6.00 D x (distance from optical center, in cm) = 5^{Δ}

Solving for the amount of decentration needed, we get 5/6 cm, or about 8mm.

6. a. - 4.50 - 1.50 x 180°
   b. + 0.50 - 3.00 x 90°
   c. + 1.25 - 1.25 x 135°
   d. + 3.50 - 1.25 x 90°

7.

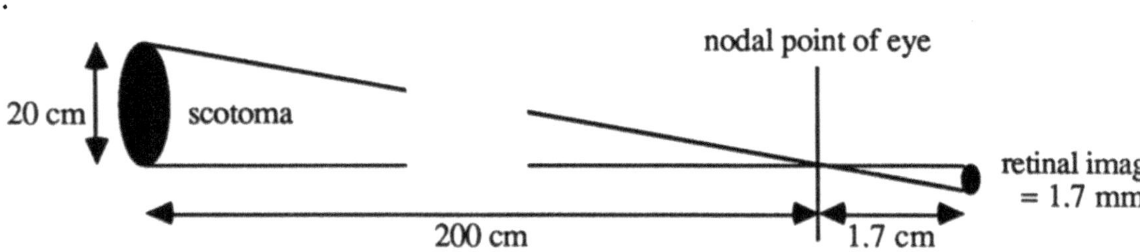

8. The magnification needed for a 20/40 line to appear the size of a 20/200 line is 5x. The magnification provided by a Galilean telescope is given as $M = - P_2 / P_1$,where P_2 is the power of the eyepiece and P_1 is the power of the objective. In this case we know the magnification required and the power of the eyepiece (P_2). Solving for the needed objective power, we get

$$5x = -(-20)/ P_1 \Rightarrow P_1 = \text{objective power} = \mathbf{+4}$$

Multiple-Choice Questions

1. A +5.00D hyperope, with a PD (inter-pupillary distance) of 5 centimeters, is mistakenly given glasses which are decentered outward by 5mm OU. The total amount of dioptric convergence power required to observe an object at 1/3 meter is:

   a. 10^{Δ}

   b. 15^{Δ}

   c. 20^{Δ}

   d. 25^{Δ}

   e. 30^{Δ}

2. You perform a refraction and plan to prescribe -1.00D sphere OD and -3.00D sphere OS with a +2.00D add bilaterally. Assuming a distance to the optical center of 8mm, the best way to treat the induced vertical phoria in the left eye is:

   a. Slab-off 1.5^{Δ} OS.

   b. Add a fresnel paste-on prism.

   c. Prescribe a flat-top segment OD and a round top segment OS.

   d. Prescribe separate spectacles for distance and reading.

   e. Decenter the left lens up 8mm.

3. Which of the following is true of a Galilean telescope?

   a. The primary focal length of the plus lens is equal to the secondary focal length of the minus lens

   b. The secondary focal length of the plus lens is equal to the primary focal length of the minus lens

   c. The minus lens has greater power than the plus lens

   d. None of the above

4. Keratometry readings of a cornea are 44.00D @ 90° and 42.00D @ 180°. The manifest refraction is -6.00 + 4.00 x 90°. What is the correction for lenticular astigmatism?

   a. +4.00 x 90°

   b. +4.00 x 180°

   c. +2.00 x 90°

   d. +2.00 x 180°

   e. +6.00 x 90°

5. A 24 year old patient can be corrected to 20/20 OU with a -0.50 +3.00 x 180° in each eye. Which Rx is the best if you want to reduce the cylindrical power by one diopter?

   a. -1.00 + 2.00 x 180°

   b. -2.00 + 2.00 x 180°

   c. plano + 2.00 x 180°

   d. -0.50 + 2.00 x 90°

6. A planoconvex lens made of crown glass has a power of +5.00D. The index of refraction is 1.5. What is the power of this lens in water (index of refraction 1.33)?

   a. 1.67

   b. 2

   c. 5

   d. 6

7. A patient observes an object 5 meters away and a 5^{Δ} base out prism is interposed in front of the observing right eye. The image of the object will be displaced:

   a. 25 cm to the left

   b. 25 cm to the right

   c. 50 cm to the left

   d. 50 cm to the right

   e. None of the above

8. A monocular 40 year old myope with an accommodative amplitude of 5.00D has a near point of accommodation at 4 centimeters without his eyeglasses. When he uses his -14.00D contact lenses, his near point of accommodation is:

   a. 9.1 cm

   b. 10 cm

   c. 11 cm

   d. 20 cm

   e. 30 cm

9. A point source of light, 1/3 of a meter to the left of a -7.00 D lens will have its image where?

   a. 10 cm to the right of the lens

   b. 10 cm to the left of the lens

   c. 25 cm to the right of the lens

   d. 25 cm to the left of the lens

10. In problem 9 above, what is the magnification of the optical system?
    a. Erect and minified by 1/3
    b. Inverted and minified by 1/3
    c. Erect and magnified by 1/3
    d. Inverted and magnified by 1/3

11. A patient uses a 12 D lens as a simple magnifying glass. What is the resultant magnification?
    a. 2x
    b. 3x
    c. 4x
    d. 6x
    e. 9x

12. You are given a spherical lens of unknown power. With the assistance of a friend holding a meter stick upright 10 meters away, you determine that the lens displaces the image of the meter stick 5 cm when you look through a point 15 mm below the optical center of the lens. What is the power of this lens?
    a. 5 D
    b. 3 D
    c. 0.33 D
    d. 1 D
    e. 2.5 D

13. A Galilean telescope has a +4 D objective and a -20 D eyepiece. What is the power and direction of the image?
    a. An inverted image, magnified 5x
    b. An erect image, magnified 5x
    c. An inverted image, magnified 80x
    d. An erect image, magnified 80x

14. What power telescope would be necessary to allow a patient with 20/200 acuity read the 20/40 line on a Snellen chart 6 meters away?
    a. 2x
    b. 2.2x
    c. 4x
    d. 5x
    e. 6x

15. Your manifest refraction of a myopic patient is -8.00 + 2.00 x 90°, vertex distance 10 mm. Keratometry readings are 44.00 @ 90° and 42.00 @ 180°. You plan to fit a rigid gas permeable contact lens 1 D steeper than the "flattest K". To compensate for this 1 diopter difference as well as for the vertex distance, what should the approximate power of the lens ordered be?

    a. -5.50 D
    b. -6.12 D
    c. -6.50 D
    d. -7.50 D
    e. -8.25 D

16. A medical student comes to you for a refraction. He tells you that he remembers his refraction is +7.50 D or +7.00 D, but he isn't sure which is for his contact lens and which is for his spectacles. Which of these is more likely the power of the **contact lens** and would this give a relatively larger or smaller image when compared with the spectacle?

    a. +7.00 D; larger image
    b. +7.00 D; smaller image
    c. +7.50 D; larger image
    d. +7.50 D; smaller image
    e. it is impossible to determine from the data given

17. You prescribe a spectacle correction of -3.00 + 2.00 x 90° to a patient. Which of the following describe the optical characteristics of this patient's eye?

    a. +1.00 + 2.00 x 180°
    b. -3.00 + 2.00 x 90°
    c. +2.00 - 3.00 x 90°
    d. +5.00 - 2.00 x 90°
    e. +3.00 - 2.00 x 180°

18. A patient wears a spectacle correction of -2.00 -1.00 x 90° OD and -4.00 -1.00 x 90° OS, with the optical centers properly aligned. She has her lenses fitted into a new frame; in cutting the lenses to fit, a careless optician decenters both lenses outwards by 5 mm. What is the total amount of induced prism?

    a. 2.5Δ base out

    b. 2.5Δ base in

    c. 4.0Δ base out

    d. 4.0Δ base in

    e. 4.5Δ base out

19. A -3.00 D myope is completely cyclopleged. In order to see clearly an object 25 cm in front of his eye, what power contact lens must you place on his cornea?

    a. +1.00 D

    b. +2.00 D

    c. +3.00 D

    d. +4.00 D

    e. +5.00 D

20. In designing bifocals, "image jump" can best be minimized by:

    a. using a small bifocal segment

    b. using a blended bifocal segment having no visible line of separation

    c. lowering the bifocal segment by 3 mm

    d. using a bifocal type that has the segment's optical center near the segment top

    e. placing the top of the segment as close as possible to the optical center of the distance segment

21. A patient with the distance correction +3.00 + 1.00 x 180° OD and Plano + 5.00 x 90° OS has become presbyopic, and you order a +3.00 flat-top executive-style bifocal segment (*i.e.*, the optical centers of the near and distance correction are the same) added to his current correction. What difference in vertical prism between the two eyes will be induced in this patient when he reads 9 mm below the optical center of the spectacles?

    a. 1.8Δ

    b. 2.7Δ

    c. 3.6Δ

    d. 4.5Δ

    e. 5.0Δ

22. You add a small stick-on mirror to the driver's-side mirror on your car to eliminate your "blind spot". It has a radius of curvature of 20 cm. The mirror is 1/3 meter from the driver. What is the reflecting power of this mirror?

    a. -5 D

    b. +5 D

    c. -10 D

    d. -20 D

23. Relative to the image seen in the flat mirror it is mounted on, the image reflected by the stick-on mirror in problem 22 will appear to be:

    a. 1/2 the size and upright

    b. 1/4 the size and upright

    c. 1/2 the size and inverted

    d. 1/4 the size and inverted

    e. 1/8 the size and upright

24. What is the reflecting power of a cornea with a radius of curvature of 8 mm?

    a. -250 D

    b. -300 D

    c. -350 D

    d. -400 D

Multiple-Choice Answers

1. c
2. a
3. d
4. c
5. c
6. a
7. a
8. a
9. b
10. a
11. b
12. c
13. b
14. d
15. c
16. d
17. e
18. d
19. a
20. d
21. a
22. c
23. b
24. a

Crib Sheet

The following crib sheet contains most of the formuli, conventions, and definitions required to manipulate the majority of optical problems that challenge the student of optics and ophthalmology. Consider this to be only the briefest of reviews of a brief review of optics.

c (speed of light) = $\lambda\upsilon$

•

short λ (blue) refracts more

•

λ changes with the medium

•

Snellen 20/20 "E" subtends 5 minutes arc and is 0.9 cm high

•

U + P = V

U = object • P = lens power • V = image
U = 1/u in meters • V = 1/v in meters
F_1 = object point where V = 0 (parallel input)
F_2 = image point where U = 0

•

Magnification (M) = U/V

a minus value means the image is *inverted*
a value < 1 means *minification*
For a "+" lens, M = D/4
For a Galilean Telescope, M = $-P_2/P_1$

•

Power of a Sphere (P_s) = (N' - N)/r

N' and N are indices of refraction
Light travels from N to N'
r = radius of curvature, in meters

•

Power of a Mirror (P_m) = -2/r

r = radius of curvature, in meters
"r" of a convex mirror has a positive value • "r" of a concave mirror has a negative value

•

•

Snell's Law

$N\sin i = N'\sin i'$ (or simply $Ni = N'i'$)

Power of axis perpendicular to shorthand meridian

•

Prism diopter (Δ) = centimeters displaced/meter

$1^{\Delta} = 1.7°$ (approximately)

•

Common indices of refraction (N)

Vacuum = 1.0 (by definition) • Air = 1.0 (by convention)

Water, Aqueous = 1.33 • Cornea = 1.38 • Lens = 1.42 • Crown Glass = 1.52

•

The Prentice Rule

Induced prism (Δ) = power of lens x cm from optical center

Flat-top Bifocal Lens gives smallest image "jump"

•

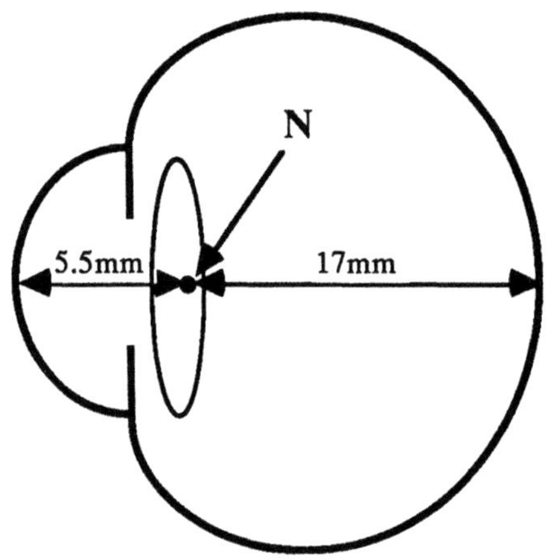

Simplified Schematic of Gullstrand's Model of the Human Eye

Index of Refraction (n) = 1.35

P = +60 Diopters

Axial Length = 22.5mm

•

Accommodation

Range of Accommodation is given in meters

Amplitude of Accomodation is given in diopters

Accommodation required with magnification = M^2 x (original accommodation)

•

GPSR Compliance

The European Union's (EU) General Product Safety Regulation (GPSR) is a set of rules that requires consumer products to be safe and our obligations to ensure this.

If you have any concerns about our products, you can contact us on ProductSafety@springernature.com

In case Publisher is established outside the EU, the EU authorized representative is:

Springer Nature Customer Service Center GmbH
Europaplatz 3
69115 Heidelberg, Germany

Batch number: 10692442

Printed by Printforce, the Netherlands